Oxford Chemistry Series

General Editors

P. W. ATKINS   J. S. E. HOLKER   A. K. HOLLIDAY

Oxford Chemistry Series

Basic chemical thermodynamics

Fourth Edition

E. BRIAN SMITH

Master of St. Catherine's College,
Oxford

CLARENDON PRESS · OXFORD
1990

PHYSICS

Oxford University Press, Walton Street, Oxford OX2 6DP

Oxford New York Toronto
Petaling Jaya Singapore Hong Kong Tokyo
Delhi Bombay Calcutta Madras Karachi
Nairobi Dar es Salaam Cape Town
Melbourne Auckland

and associated companies in
Berlin Ibadan

Oxford is a trade mark of Oxford University Press

Published in the United States
by Oxford University Press, New York

© *E. Brian Smith 1973, 1977, 1982, 1990*

First published 1973
Second edition 1977
Third edition 1982
Reprinted (with corrections) 1983
Reprinted 1984, 1986, 1987, 1988, 1989
Fourth edition 1990

British Library Cataloguing in Publication Data
Smith, E. Brian (Eric Brian), 1933–
Basic chemical thermodynamics,—4th ed.
1. Chemical reactions. Thermodynamics
I. Title
541.369
ISBN 0–19–855565–2
ISBN 0–19–855564–4 pbk

Library of Congress Cataloging-in-Publication Data
Smith, E. Brian (Eric Brian), 1933–
Basic chemical thermodynamics/E. Brian Smith.—4th ed.
p.   cm.—(Oxford chemistry series; 35)
Includes bibliographical references.
1. Thermodynamics.   I. Title.   II. Series.
QD504.S57   1990   541.3'69—dc20   90-6933
ISBN 0–19–855565–2
ISBN 0–19–855564–4 (pbk.)

First Spanish Edition 1977
First Portugese Edition 1977
First Chinese Edition 1988
First Dutch Edition 1989
First Polish Edition 1990

Typeset by Macmillan India Ltd, Bangalore 560 025
Printed in Great Britain
by Bookcraft (Bath) Ltd
Midsomer Norton, Avon

Editor's foreword

Thermodynamics is a remarkable intellectual structure, for it deals with the mathematical relations between observations, and is independent of theoretical models of the microscopic nature of matter. And because of this independence and immediacy it is also remarkably useful. From diverse experiments (and so now from tabulated data) we can predict properties such as the direction of spontaneous chemical and physical change and the composition of reaction mixtures at equilibrium, and the response of these properties to changes in the conditions. We can understand thermodynamics more deeply if we can relate its ideas to the underlying molecular properties; but such a step is not essential, and it is possible to use its results (and to deduce more results) so long as its techniques are understood. This volume therefore sets thermodynamics in a chemical context; it is not exhaustive, but it surveys in a brief and simple fashion the ideas of chemical thermodynamics. It emphasizes some of the most useful results of thermodynamics in a businesslike way, and illustrates its most common applications.

P.W.A.

Preface to the fourth edition

Since the First Edition the most significant change to the book has been the addition of a new chapter which provides a simple treatment of the molecular basis of thermodynamics. Though this chapter has been placed at the end of the book it has been written in such a way that it could be employed with advantage at an earlier stage of a first course in chemical thermodynamics. Indeed, a prompt introduction to the elements of statistical thermodynamics can be very helpful in reinforcing the fundamental concepts of classical thermodynamics.

For this edition I have again tried to compromise between accommodating the views of teachers of the subject and those of their students, which more often than not appear to conflict. Most often I have favoured the views of the students who in general advocate less radical surgery. The result is a little more rigour in specifying the dimensions within logarithmic expressions, the addition of more worked examples in the text, a reference to Ellingham diagrams and a number of minor revisions.

Physical Chemistry Laboratory,   E.B.S.
Oxford
1990

Acknowledgements

Over four editions of a book one becomes indebted to many for their advice and help. In addition to those named in the Preface to the first edition I would like to thank: Professor J. S. Rowlinson, Drs M. L. Khan, G. P. Matthews, G. D. Meakins, M. Rigby, M. Spiro, A. R. Tindell, and B. H. Wells, Miss Anne Buckley, Miss Emma Collingwood, and Mr C. D. Eley. I am indebted to Dr. J. van Mourik and Bohn, Scheltema and Holkema publishers for the use of problems from the Dutch edition.

Preface to the first edition

The first time I heard about Chemical Thermodynamics was when a second-year undergraduate brought me the news early in my freshman year. He told a spine-chilling story of endless lectures with almost three hundred numbered equations, all of which, it appeared, had to be committed to memory and reproduced in exactly the same form in subsequent examinations. Not only did these equations contain all the normal algebraic symbols but in addition they were liberally sprinkled with stars, daggers, and circles so as to stretch even the most powerful of minds.

Few would wish to deny the mind-improving and indeed character-building qualities of such a subject! However, many young chemists have more urgent pressures on their time. Chemical thermodynamics need not be a particularly esoteric branch of algebra: only a handful of thermodynamic relations are important for the student when he first meets the subject. It is essentially a practical subject that interrelates quantities which can be measured in the laboratory (some more easily than others).

This book is not intended to be a formal textbook of thermodynamics. It is intended to give the beginner some familiarity with the concepts of thermodynamics and a knowledge of the thermodynamic relations he will use in the laboratory. At a recent conference on the teaching of thermodynamics it was concluded that there was 'no place for axiomatics for introductory students'.†
The presentation in this book is certainly non-axiomatic and occasionally non-rigorous. There is often a direct conflict between rigour and clarity in the presentation of elementary thermodynamics, and this book strays from the path of rigour more than most. The analogies used to provide 'insight' are (like all analogies) capable of being misleading if they are examined too closely or pursued too far. Nevertheless teaching the subject over a number of years has convinced me that this approach is on the whole beneficial.

The book assumes that the reader will have taken a course in elementary physics and have some (passing) acquaintance with the concepts of potential and kinetic energy, work, heat, temperature, and the perfect-gas state. A knowledge of very elementary calculus is also assumed.

Chapters 1–3 outline the basic concepts of chemical thermodynamics: energy, entropy, and equilibrium. Chapter 4 introduces free energy and

† See H. A. Skinner (1971) *Chemistry in Britain*, 7, 438.

develops the thermodynamic approach to the understanding of equilibrium in chemical systems. The order of presentation of material differs from that most commonly employed in that the determination of thermodynamic quantities is deferred until Chapter 5. In this chapter the changes in free energy and entropy accompanying chemical reactions are treated together with enthalpy changes, which are more often tackled at an earlier stage. These five chapters would form the basis of a suitable introductory course in Chemical Thermodynamics. Chapter 6 develops the concept of the ideal solution and applies it to colligative properties. Much of this chapter could also be included in a first approach to the subject. Chapter 8 and much of Chapter 7, on the other hand, are essentially notes intended to serve as a bridge between the elementary thermodynamics of the earlier chapters and more complete treatments.

The notation is consistent with the IUPAC recommendations of 1969 and SI units are employed. Not a great deal of thermodynamic reference data is available in SI units and therefore a reasonable quantity of such data is provided in Appendices 1 and 2. Stress has been laid throughout on the physical principles underlying the subject, and extensive mathematical manipulation of equations has, so far as is possible, been avoided. In order to emphasize that chemical thermodynamics is not an exercise in elementary algebra the individual equations have not been numbered. The more important relations have been identified in the text. When a reference to the origin of an equation is necessary the section of the book in which it is introduced is given.

I am indebted to many teachers and colleagues, not least to Professor J. H. Hildebrand who at 90 is still contributing to my education in thermodynamics. I hope other writers on the subject from whom I have drawn ideas or analogies will interpret this as a compliment. I would like to thank Dr G. C. Maitland, Dr. R. P. H. Gasser, Mr P. Scott and Dr L. A. K. Staveley who contributed numerous ideas for improving the manuscript and Dr P. W. Atkins for his editorial and scientific advice.

There will be, no doubt, all too many errors and weak arguments left for the reader to discover. I shall be very grateful to be informed of them.

Physical Chemistry Laboratory,   E.B.S.
Oxford
1972

Contents

Notation

The notation for thermodynamic quantities used in this book follows the recommendations of the International Union of Pure and Applied Chemistry as published in the volume IUPAC *Manual of Symbols and Terminology for Physicochemical Quantities and Units* (Butterworths, London, 1969).

Symbols

a	activity	q	heat
A	Helmholtz free energy	Q	charge
c	concentration	S	entropy
C	heat capacity	T	temperature
E	electromotive force	U	internal energy
f	fugacity	v	velocity
G	Gibbs free energy	V	volume
H	enthalpy	w	work
K	equilibrium constant	W	number of microstates
m	molality, molecular mass	x	mole fraction
M	mass, relative molecular mass (molecular weight)	z	partition function
		ε	molecular energy
n	amount of substance	μ	chemical potential
p	probability	ξ	extent of reaction
P	pressure		

Where, on occasions, these symbols have been used to represent different quantities this is made clear in the text.

In order to simplify the equations, properties of the whole system and properties per mole have not been distinguished by changes in notation. The context in which a property is defined is made clear in the text. In most cases the thermodynamic properties refer to one mole of material and changes in thermodynamic properties are for one mole of reaction.

The physical state of a substance is indicated by symbols in parentheses following the symbol for the relevant property. Thus $H_i(\mathrm{l})$ would indicate the enthalpy of substance i in the liquid state. Thermodynamic processes are indicated by subscripts. Thus ΔH_{vap} indicates an enthalpy change on vaporization. It will refer to the change for one mole of substance unless otherwise stated. The melting point of a substance would be denoted T_{fus}.

State		Process	
gas	(g)	vaporization, boiling	vap
liquid	(l)	fusion, melting	fus
solid	(s)	mixing	mix
solution	(soln)	transition	trans
aqueous solution	(aq)	formation from elements	f

Standard states. It is convenient, when writing the equations of chemical thermodynamics to define a number of *standard states*. We have, in general, identified the thermodynamic functions of substances in a standard state by a superscript $\ominus$, for example $\mu^{\ominus}$. One specific standard state that is used throughout this work is that of a pure substance in its normal state of matter at 1 atmosphere pressure. This we identify with the symbol 0 as in μ^0. A further symbol, *, is used to indicate a pure substance at a pressure that is not necessarily 1 atmosphere. To summarize:

0   Substance at 1 atm and in the pure state,

*   Substance at arbitrary pressure and in the pure state,

$\ominus$ Standard state in general (often hypothetical).

The chemical components of a system are indicated by subscripts as in $\Delta U = U_B - U_A$. The subscript i is used to denote an unspecified chemical compound in equations of general applicability.

Subscripts A and B are also used to indicate different states of the system. Very occasionally other subscripts are used and these are defined within the section in which they appear, for example x_{id}, the ideal solubility of a solid in a liquid, referred to in Section 7.2.

1
Introduction

1.1 The scope and nature of chemical thermodynamics

Thermodynamics is one of the most powerful techniques we can use in the study of natural phenomena. Despite its name it is not concerned with the dynamics of systems but rather their equilibrium positions, those positions in which they show no tendency to further change. It requires no assumptions about the nature of the molecules which make up a system, nor even is it necessary to assume that molecules exist, and consequently its conclusions are quite general. Thermodynamics gives scientists a set of relations between *macroscopic* properties that we can measure in the laboratory, such as temperature, equilibrium constant, volume, and solubility. These relations can all be derived from a few initial postulates, the so-called Laws of Thermodynamics.

Consider an important example of chemical equilibrium:

$$N_2 + 3H_2 \rightleftharpoons 2NH_3.$$

This is the basis of the Haber Process for ammonia.† As chemists we would like to be able to answer a number of questions about such equilibria.

(i) What properties of N_2, H_2, and NH_3 determine the position of equilibrium at a particular temperature and pressure? In other words from what we know, or could find out about the properties of these gases, we would like to predict how much NH_3 will be formed under a given set of conditions.

(ii) To what extent will the position of equilibrium be altered if we alter temperature and pressure?

These are the questions (among many others) to which thermodynamics gives an answer. However, the equilibrium positions predicted by thermodynamics may not always be attainable in practice. Indeed, in our example, the synthesis of ammonia, a catalyst is essential to facilitate the attainment of equilibrium.

---

† The discovery of this process by the application of thermodynamic arguments is said to have saved Germany from almost immediate defeat in the First World War by providing a supply of nitrates for the manufacture of explosives—an example of the relevance of the subject.

1.2 Equilibrium in mechanical systems

Experience with the physical world gives us good insight into the position of equilibrium in mechanical systems. We shall use the word system to mean that portion of the universe that we have under investigation at any particular time. The system is separated from the rest of the universe (the surroundings) by boundaries which may be physical like the walls of a container or less concrete as in the example below.

System doing no work

Let us first turn our attention to a system that does no work. Later we shall have to define what we mean by work, but for the moment we can regard work as the lifting of a weight. An example of a system that does no work would be a ball rolling to the bottom of a bowl or a toboggan on a hillside (Fig. 1.1). We know that such objects will move, if free to do so, to a position of minimum potential energy.† We are also aware that as the object moves downhill we *could* use it to do work, for example pull another lighter object uphill by means of a rope and pulley. As our toboggan moves down to the bottom of the hill it loses the capacity to do work, which in this simple mechanical system is just its potential energy $U = Mgh$, where M is its mass, g the acceleration due to gravity and h its height above its equilibrium position. At its equilibrium position its potential energy will be minimum and so $dU = 0$. Furthermore, any *spontaneous* movement of the toboggan on the hillside will be such as to reduce its potential energy and therefore to reduce its capacity for doing work—a toboggan never spontaneously starts sliding uphill.

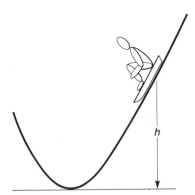

Fig. 1.1. Approach to equilibrium for a mechanical system that does no work.

† This statement implies the presence of frictional forces. In their absence the toboggan would oscillate about the position of minimum potential energy for ever!

System doing work

Now let us look at a system that can do external work. We shall again regard the toboggan on the hillside as our system but now it is coupled so that as it slides downhill it can raise a weight (Fig. 1.2). It is perhaps easier to understand factors controlling equilibrium in this situation if we simplify our model still further and consider two weights hanging over a frictionless pulley (Fig. 1.3). The system will be at rest when $M_1 = M_2$ as the forces will then be balanced.

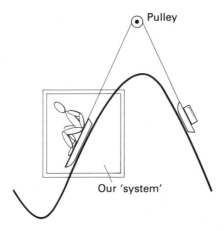

Fig. 1.2. Equilibrium for a mechanical system harnessed to do work.

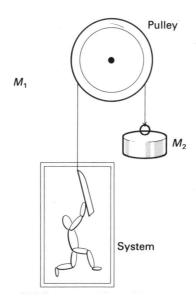

Fig. 1.3. A mechanical system harnessed to do work.

We can define this position of equilibrium in another way that will be useful in our study of thermodynamics. If we allow the mass M_1 to fall through a distance dh it will do work $M_2 g\ dh$, by lifting M_2 (where $M_2 < M_1$). If we make M_2 closer in magnitude to M_1 we shall get correspondingly more work, until at $M_1 = M_2$ an infinitesimal displacement will lead to work $M_1 g\ dh$. This is the *maximum work* M_1 can do, as it is now lifting an equal mass. When $M_1 = M_2$ the system is of course at equilibrium; therefore we can define the equilibrium condition of the system as that for which a small displacement leads to the system doing the maximum possible work.

The work done by such a system is equal to the loss of potential energy, $-dU$, of the system. As we have established in the previous section, if the system is arranged so as to be incapable of doing external work, then at equilibrium, as the work done by the system is equal to $-dU$, we have $dU = 0$ and the potential energy will be a minimum.

1.3 Reversibility and equilibrium

The processes we observe in nature are *irreversible*. Our toboggan sliding downhill dissipates its potential energy as frictional heat. To restore the vehicle to the top of the hill we must do work—it will not return spontaneously. Furthermore we cannot, for reasons to be explained later, collect the heat generated by the toboggan and use it (in an engine) to generate enough work to restore the situation.

However it is possible to imagine *reversible* processes. In Fig. 1.4 if $M_1 = M_2 + \Delta M$ the heavier weight will fall with increasing speed. If it falls through a distance h the minimum work required to restore the original situation will be $\Delta M g h$. However, if $M_1 = M_2 + dM$, so that the weights differ by only an infinitesimal amount, the fall of M_1 would be infinitely slow and will do the maximum amount of work. Only an infinitesimal quantity of work, $dM g h$ would be required to restore the system, that is the weight M_1, to its starting point. At every point during the reversible process

$$M_1 = M_2 + dM$$

a condition which is virtually indistinguishable from $M_1 = M_2$, the condition for equilibrium in the system. A reversible change is therefore one conducted so that the system is always at equilibrium and is consequently, unlike natural observable processes, infinitely slow.

To summarize—if two opposing forces act on a body, as illustrated in Fig. 1.5, we may identify conditions that will lead to spontaneous, irreversible changes and others that will lead to reversible changes. For a reversible change $F_1 = F_2 + dF$; such a change will be infinitely slow but will do the maximum amount of work. A spontaneous observable change requires

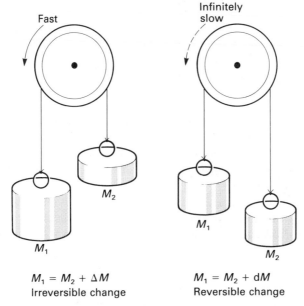

$$M_1 = M_2 + \Delta M$$
Irreversible change

$$M_1 = M_2 + dM$$
Reversible change

Fig. 1.4. Irreversible and reversible changes in a mechanical system.

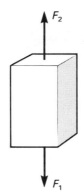

Fig. 1.5. Two opposing forces acting on a body (see text).

$F_1 = F_2 + \Delta F$; it will proceed at a discernible rate but do less than the maximum possible work.

The conditions that must be satisfied for a reversible change are the same as the conditions that must be satisfied for the system to be in equilibrium. Even though it appears to contribute little to our understanding of equilibrium in mechanical systems, the idea of reversible processes as the limiting behaviour of observable processes is of great importance in the study of equilibrium in chemical systems.

1.4 Why we need thermodynamics

Having disposed of equilibrium in mechanical systems so expeditiously you might wonder why thermodynamics takes up so much time in the teaching of chemistry. The reason is that the rules we have established for mechanical systems are not all satisfactory in the world of physico-chemical phenomena. When the block in the dish illustrated in Fig. 1.6 slides to its equilibrium position the excess potential energy is given out as heat (unless the block is harnessed to lift a weight and do work). If chemical processes followed similar rules we would expect them to liberate heat as they approached equilibrium.† The liberated heat would cause the system to warm up. If we add some solid NaOH to water this is indeed what we observe. The equilibrium position—a solution of NaOH in water—has lower energy, as energy in the form of heat is liberated (Fig. 1.7). Such observations led the pioneers of thermochemistry (as this area of investigation is called) to suggest that all spontaneous chemical reactions should be accompanied by the evolution of heat (i.e. be *exothermic*).

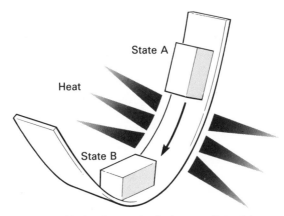

Fig. 1.6. Approach to equilibrium in a mechanical system. Potential energy is dissipated as the block moves to a state of lower energy.

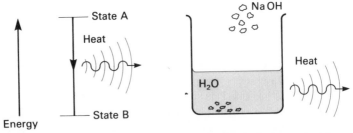

Fig. 1.7. Approach to equilibrium in an exothermic chemical process: heat is evolved.

† This is true only if the chemical processes do no work.

Thomsen (1854) and Berthelot thought that heat changes could be used to explain the directions of chemical reactions.

However, if we add solid $NaNO_3$ to water we find that heat is absorbed and the system cools down. (The process is *endothermic*.) In this case we have a system which 'climbs uphill' on the energy scale to reach its position of equilibrium (Fig. 1.8). In other words in this system energy cannot be the sole factor determining the position of equilibrium.

The existence of another force driving physico-chemical systems to equilibrium can be further illustrated if we consider systems whose energy is constant.

1. *Expansion of a gas.* Consider a gas confined to one of two bulbs connected by a stopcock, the other bulb being evacuated (Fig. 1.9). If the stopcock is opened the gas will flow so as to distribute itself uniformly between the two vessels. For a perfect gas (and most real gases are almost perfect under normal conditions) there is no change in energy accompanying this expansion. Nevertheless, there is clearly *some* driving force causing the gas to distribute itself between the two vessels.

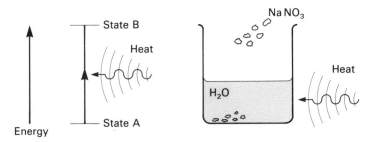

Fig. 1.8. Approach to equilibrium in an endothermic process: heat is absorbed.

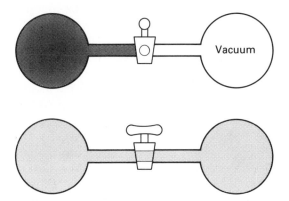

Fig. 1.9. Expansion of a gas into a vacuum.

2. *Flow of heat*. If a block of hot metal is placed in thermal contact with a colder block, energy, in the form of heat, will flow until both bodies are at the same temperature (Fig. 1.10). If such a system is insulated from its surroundings there will be no change in the total energy. Again, a property other than energy must determine the approach to equilibrium.

The existence of this extra factor explains why, in order to understand equilibrium in physico-chemical systems, we require a more sophisticated analysis than sufficed for mechanical systems.

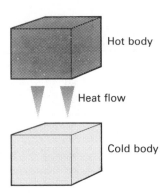

Fig. 1.10. Flow of heat from a hot body to a cold body.

1.5 The mole

When performing calculations in chemical thermodynamics, and indeed in many other areas of chemistry, we need a convenient unit to represent quantity of material. We could calculate the value of properties per molecule but these would be very small and not related to the magnitudes we measure in the laboratory. To solve this problem we use the concept of the mole.

The mole is the fundamental unit of quantity of material. It provides a convenient way of scaling up molecular masses to those on a laboratory scale of measurement. A mole of substance is equal to as many molecules of that substance as there are atoms in exactly 12 g of the ^{12}C isotope of carbon. This number, called Avogadro's constant, is 6.022×10^{23} mol^{-1}. We can have moles not just of molecules but of ions, atoms, or any other particles.

The mass of a mole of molecules of a substance is termed its molecular mass and is usually expressed in units of g mol^{-1}.

The molecular weight of a substance is more properly termed the relative molecular mass. This is a dimensionless ratio of the mass of a molecule to a unit of mass which is *very approximately* the mass of a hydrogen atom. This unit, the atomic mass unit (a.m.u.), is more accurately defined as $\frac{1}{12}$ the mass of an atom of ^{12}C and equals 1.6605×10^{-27} kg.

1.6 The perfect gas

We must first consider the properties of the perfect-gas state. This state, to which real gases tend at low pressures, plays an important role in chemical thermodynamics.

A perfect gas may be defined as one that satisfies two conditions:

1. The pressure, volume, and the absolute temperature are related by the equation

$$PV = nRT$$

where n is the number of moles of substance and R is the gas constant.

2. The energy of the gas depends only on its temperature and not on its pressure or volume. (It is possible to show that this second condition is a direct consequence of the first. This will be done later—for the moment we shall accept the second requirement as an independent one.)

Though we do not need to invoke a knowledge of the molecular nature of matter in order to understand the conclusions of chemical thermodynamics, it is helpful to do so in the case of the perfect gas. The perfect-gas state occurs when the molecules behave simply as mass points which do not interact; that is they neither repel nor attract each other. The total energy of such a gas is just its kinetic energy and this may be shown to be directly proportional to the absolute temperature. As there is no potential energy arising from the forces between the molecules it is clear that the energy of a perfect gas will not change when its volume is varied and the average distance between molecules altered.

In a mixture of perfect gases, since the molecules take up no space and do not interact, each gas behaves as if it were alone in the container. The total pressure is therefore just the sum of the pressures that each of the gases would exert if it were alone in the same volume. These pressures are called the *partial pressures* of the gases. If n_A molecules of perfect gas A and n_B molecules of perfect gas B are mixed we may write the total pressure P as $P = P_A + P_B$. The contributions of A and B to the total pressure will depend simply on the number of moles of each of these substances present. Thus the partial pressures P_A and P_B are given by

$$P_A = \left(\frac{n_A}{n_A + n_B}\right)P \quad \text{and} \quad P_B = \left(\frac{n_B}{n_A + n_B}\right)P.$$

$\{n_A/(n_A + n_B)\}$ is called the *mole fraction* of substance A in the mixture, and is usually written x_A.

2
Energy

2.1 Work

We have already made a distinction between two types of energy which may be transferred from one system to another, work and heat. Indeed if energy is applied to one's toe by dropping a weight on it or by putting it in hot water the distinction is readily appreciated. In thermodynamics the formal definition of work—'Work is the transfer of energy from one mechanical system to another. It is always completely convertible to the lifting of a weight'—is best illustrated by example.

Work can be expressed in terms of a force and the displacement of its point of action:

$$w = \int_{L_1}^{L_2} F \, dL.$$

Expansion is an example of work which frequently occurs in chemical problems. This is the work done pushing back the atmosphere when a system changes its volume. We shall define work done *on* the system as positive (as this leads to the system gaining energy) and work done *by* the system as negative. If a gas expands against an external pressure P_{ex} (Fig. 2.1), the work involved is

$$w = \int_{L_1}^{L_2} (-P_{ex} A) \, dL$$

where A is the area of the piston and L the distance it moves. As $A \, dL = dV$

$$w = - \int_{V_1}^{V_2} P_{ex} \, dV.$$

If the external pressure is continuously adjusted so that it is kept equal to P, the pressure of the gas within the cylinder, the system is always at equilibrium and the expansion is reversible. Under these conditions

$$w_{rev} = - \int_{V_1}^{V_2} P \, dV.$$

The negative sign occurs because we consider work done *by* the system to be a negative quantity as it is energy 'lost' from the system. For an expansion,

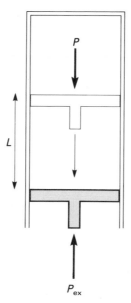

Fig. 2.1. Expansion of a gas against an external pressure.

ΔV is a positive quantity and the system does work leading to a negative value of w.†

Example

Calculate the work done when 1 mole of gas expands from 5 dm³ to 10 dm³ against a constant pressure of 2 atm.

For constant pressure the work

$$w = P\Delta V$$

Converting the data in the example to SI units

$$w = 2 \times 1.013 \times 10^5 (10^{-2} - 5 \times 10^{-3}) \text{Pa} \, \text{m}^3$$
$$w = \underline{1013} \text{ J}.$$

There are of course many other types of work. One commonly occurring in physical chemistry is electrical work in which a charge Q is moved through an electric potential difference E (not to be confused with energy). In this case $w = - \int E \, dQ$. This is the work which can be obtained from electrochemical cells.

---

† This definition, which means that work done *by* the system is negative work, will tend to make some explanations that follow rather difficult. It is hallowed by international agreement on thermodynamic nomenclature and is therefore followed in this book.

Power is the rate at which work is done and is measured in joules per second, i.e. watts.

2.2 Heat and temperature

The understanding of what is meant by heat proved a challenge to the pioneers of thermodynamics. Indeed the equivalence of heat and work as different forms of energy was not unequivocally established until comparatively recent times. In 1842 Mayer, a German doctor, stated the Law of Conservation of Energy in its modern form, including all forms of energy, among them heat. We now define heat as 'the transfer of energy that results from temperature differences'.

The difficulty of understanding heat largely arose from its confusion with the concept of temperature. Early scientists tended to believe that objects came to thermal equilibrium when they each contained an equal amount of heat per unit volume. Black, whose work was published after his death in 1799, did much to clarify the position. He showed that different substances have different heat capacities. Thermal equilibrium is established between two bodies when their *temperatures* become equal. Heat will flow until temperature gradients disappear.† Thus temperature difference provides the driving force for the flow of heat. The relation between the amount of heat transferred to a body, and the ensuing change in its temperature, depends on its heat capacity:

$$C = \frac{dq}{dT},$$

where C is the heat capacity, q the heat, and T the temperature.

2.3 Measurement of temperature

Temperature can be measured using any body having a suitable property which depends on its 'hotness'. The expansion of a suitable liquid in a glass container is the most common method of measuring temperature. Traditionally thermometers were calibrated by placing them in water boiling under 1 atmosphere pressure. This reading was assigned the value of 100° Celsius. The origin of this scale, 0°C, was defined by the temperature of melting ice. Intermediate temperatures were measured by assuming linear behaviour. Unfortunately this assumption is not exact and the temperature we record depends on the type of thermometer used. Thus a mercury in glass thermometer would not, in general, indicate the same temperature as a thermo-

---

† The observation that when two bodies are in thermal equilibrium with a third then they must be in thermal equilibrium with each other is sometimes called the Zeroth Law of Thermodynamics. It is the basis for the concept of temperature.

couple. However, in order not to get too deeply involved in the problem of proper definition of temperature,‡ we shall define temperature as measured on the perfect-gas scale. Thus for a fixed quantity of perfect gas at constant volume $T \propto P$. This relation together with the definition of the melting point of ice as 273.15 K (or more exactly the triple point of water as 273.16 K) gives us a complete scale of temperature. Thus having defined the ice point, the steam point or any other characteristic temperature may be measured using the relation

$$P_1/P_2 = T_1/T_2$$

for a perfect gas as illustrated in Fig. 2.2.

2.4   Heat and molecular motion

Though thermodynamics does not require us to invoke the molecular nature of matter it is frequently helpful to do so. It is particularly helpful in the distinction between heat and work. Work may be regarded as the energy associated with *orderly* movements of bodies, or the particles that comprise them, for example pushing back boundaries. Movements of centre of mass or

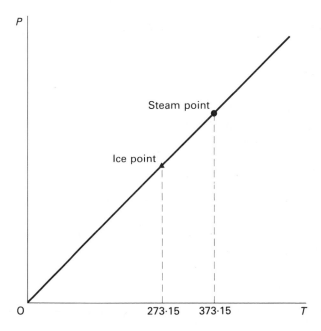

Fig. 2.2.   Temperature (T) defined in terms of the ice point (273.15 K) and the pressure of a perfect gas (P).

‡ For a fuller account of the concept of temperature see E. F. Caldin (1958). *An introduction to chemical thermodynamics.* Clarendon Press, Oxford.

flow of electrons in a wire are of this type. On the other hand heat flows are the movement of thermal energy. Thermal energy arises from the *disorderly* random motion of molecules. Such random motion cannot *all* be converted into work unless we are able to bring the molecules to rest—an impossibility under normal conditions. Only at the absolute zero of temperature could we imagine a state in which the motion of the molecules could not be further diminished and all their energy converted into work.

2.5 Conservation of energy

This principle is often referred to as the First Law of Thermodynamics: '*The algebraic sum of all energy changes in an isolated system is zero.*' An isolated system is one that cannot exchange energy or matter with its surroundings.

The First Law tells us that energy may be converted from one form to another but cannot be created or destroyed.† When a chemical system changes from one state to another the net transfer of energy to its sur-roundings must be balanced by a corresponding change in the internal energy of the system. If the system starts in state A and changes to state B we may write

$$\Delta U = U_B - U_A = q + w,$$

where U is the energy contained in the system, called the *internal energy*, q is the heat absorbed *by* the system, and w the work done *on* the system. ΔU depends only on the initial and final states of the system and not on the path taken. If the paths I and II between states A and B in Fig. 2.3 resulted in a different ΔU we would be able to make a complete cycle for which $\Delta U \neq 0$. This would mean that energy was being created or destroyed, in direct conflict with the First Law. The heat absorbed and the work done on the system may differ for the various paths, but the internal energy, their sum, must be the same.

Consider a block sliding downhill. It may do work, e.g. by raising another weight as in Fig. 2.4. If we convert all the potential energy into work by having both weights virtually equal, it will proceed downhill infinitely slowly and generate no heat.

$$\Delta U = -Mgh = +w, \qquad q = 0.$$

Or we can allow the block to do no work, as in Fig. 2.5, in which case the potential energy will be released as heat (from friction).

$$\Delta U = -Mgh = +q, \qquad w = 0.$$

† The validity of this statement rests on experimental observation. No exceptions have yet been reported. Strictly, we should include mass as a form of energy but this is not usually necessary in chemistry as significant mass changes do not occur in the course of chemical experiments.

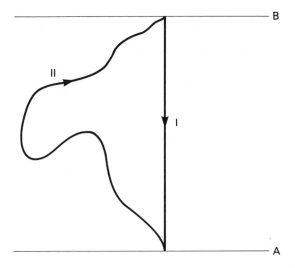

Fig. 2.3. Two possible paths (I and II) between states A and B.

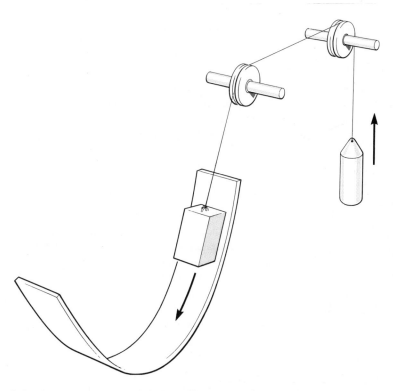

Fig. 2.4. System in which work is done. The block slides slowly down, doing work by raising the weight.

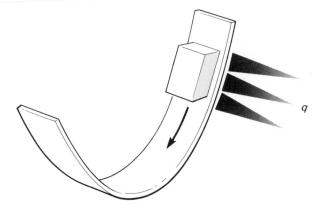

Fig. 2.5. System in which no work is done. The block slides quickly, dissipating frictional heat.

In both cases ΔU will be the same though q and w are different. All intermediate cases could occur.

2.6 State functions: a digression

The state of a perfect gas can be defined by specifying P, V, and T. As $PV = nRT$ for a fixed mass of gas we need specify only two of P, V, and T since this will be sufficient to fix the remaining variable. Indeed such is the case for any pure substance (or mixture of fixed composition)† even though it may not follow the perfect gas equation. We may write $T = f(P, V)$. Such an equation which links P, V, and T is called an *equation of state*.

We have seen that U differs from q and w in that it depends only on the state of the system. Thus if we can define the state of a substance by fixing, for instance, P and T then U will have a definite value. U, like P, T, and V, is called a *state function*.

State functions may depend on the mass of material we have: thus V and U would be twice as large if we doubled the quantity of material in the system (other things being equal). These are called *extensive* properties. On the other hand T and P are independent of the amount of material we are dealing with. These are called *intensive* properties. If we divide a system into smaller parts then the intensive properties of each portion would have the same value as for the whole system.

To illustrate the properties of state functions let us consider the two cities A and B shown in Fig. 2.6. The latitude and longitude are analogous to state functions. However we go from A to B, Δ (latitude) and Δ (longitude) will always be the same. The values will depend only on the position of A and B

† In the absence of electrical and magnetic fields, etc.

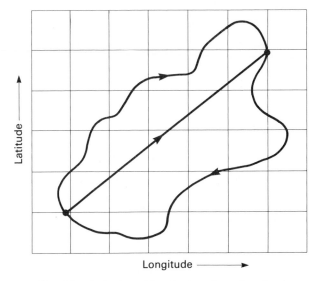

Fig. 2.6. Latitude and longitude as 'state functions'.

(i.e. the initial and final states). On the other hand the distance travelled on a journey from A to B depends on the path taken and is clearly not a state function. State functions such as U, V, and T, provide a suitable means of defining the state of a thermodynamic system in the same way as latitude and longitude define a geographical location. On any journey which starts and finishes at the same point the changes in latitude and longitude must add up to zero. So also the changes in any state functions must always add up to zero over any complete cycle.

State functions have a number of important mathematical properties.

(i) If we integrate a state function,

$$\Delta U = \int_A^B dU,$$

this integral must have a definite value which is independent of the path of integration between the limits A and B. Thus dU is said to be an exact differential.

(ii) We can write an exact differential

$$dU = \left(\frac{\partial U}{\partial x}\right)_y dx + \left(\frac{\partial U}{\partial y}\right)_x dy$$

where x and y are variables which determine the value of U, for instance any two of pressure, temperature, and volume. $\left(\dfrac{\partial U}{\partial x}\right)_y$ is the rate of change of U with changes in x when y is kept constant. It is called a partial differential coefficient. If $U = f(T, V)$ then

$$dU = \left(\frac{\partial U}{\partial T}\right)_V dT + \left(\frac{\partial U}{\partial V}\right)_T dV.$$

(iii) The order of differentiation of a state function is immaterial; thus

$$\left[\frac{\partial}{\partial V}\left(\frac{\partial U}{\partial T}\right)_V\right]_T = \left[\frac{\partial}{\partial T}\left(\frac{\partial U}{\partial V}\right)_T\right]_V.$$

If a function or its differential can be shown to satisfy any one of these conditions then the function is a state function.

Thermodynamics is largely concerned with the relations between state functions which characterize chemical systems. However, some quantities with which we shall be concerned are not state functions. For example q and w depend on the path between the states and are not state functions. dq and dw are therefore not exact differentials and it is customary to write them đq and đw to remind us that they cannot in general be integrated to give a unique value (as illustrated in Figs 2.4 and 2.5).

2.7 Enthalpy

In the laboratory chemical systems are not usually harnessed to do work. The only work they do is PV work arising from expansion or contraction. (An important exception in which chemical reactions are allowed to do additional work, other than PV work, is the electrochemical cell.) If we have a chemical system at *constant volume*, it can do no work as đ$w = -PdV = 0$. Because $dU = đq + đw$,

$$dU = (đq)_V \quad \text{and} \quad \Delta U = (q)_V$$

The increase in internal energy of the system is therefore equal to the heat absorbed at constant volume (for a system that does no work).

Most chemical experiments are carried out at constant pressure rather than at constant volume. Under such conditions the work done by the system as a result of expansion is not zero.

$$đw = -PdV \quad \text{and} \quad w = -P\Delta V.$$
$$\Delta U = q + w = U_B - U_A = (q)_P - P(V_B - V_A)$$

and

$$(q)_P = (U_B + PV_B) - (U_A + PV_A).$$

$(U + PV)$, like U, is a state function, as U, P, and V are all state functions. We call this function the *enthalpy* and it is defined by

$$H = U + PV$$

$$\Delta H = (q)_P \quad \text{and} \quad dH = (\mathrm{d}q)_P.$$

The increase of enthalpy of a system is equal to the heat absorbed at constant pressure (assuming that the system does only PV work).

Even for changes that occur at other than constant pressure ΔH has a definite value. However, under these conditions it is not equal to the heat absorbed. Similarly, ΔU has a definite value for any change, irrespective of whether it is at constant volume, but it is only for a change at constant volume that $\Delta U = q$.

The importance of this new state function, enthalpy, will become apparent when we study thermochemistry, the branch of thermodynamics concerning the heat changes associated with chemical reactions. For the moment let us note that when we see

$$CS_2 + 3O_2 \rightarrow CO_2 + 2SO_2; \quad \Delta H = -1108 \, \text{kJ mol}^{-1},$$

this means that for one mole of reaction the enthalpy of the system *decreases* by 1108 kJ and this quantity of heat will be *liberated* (at constant T and P) by the reaction. One mole of reaction is when the appropriate numbers of moles of substances (as specified by the stoichiometric coefficients) on the left hand side of the equation are converted to the substances on the right hand side of the equation. When the change in a thermodynamic property is given for a particular chemical reaction or process, it always refers to a mole of reaction unless an exception to this rule is specifically indicated.

ΔH and ΔU are usually very similar for processes involving solids and liquids, but for gases they may be significantly different. If a gas reaction involves a change of Δn moles of gases in the system then as $\Delta H = \Delta U + \Delta(PV)$, and for perfect gases $\Delta(PV) = (\Delta n)RT$,

$$\Delta H = \Delta U + \Delta n RT.$$

At 298 K $RT = 2.5 \, \text{kJ mol}^{-1}$, not a negligible quantity.

2.8 Heat capacity

We have seen (Section 2.2) that the heat capacity of a body may be defined by

$$C = \frac{\mathrm{d}q}{\mathrm{d}T}.$$

If the heat capacity is determined at constant volume (by measuring the heat required to produce a unit rise in temperature at constant V) then as

$$dU = (đq)_V \qquad \text{(Section 2.7)}$$

$$C_V = \left(\frac{\partial U}{\partial T}\right)_V.$$

If the heat capacity is measured at constant pressure, since $dH = (đq)_P$

$$C_P = \left(\frac{\partial H}{\partial T}\right)_P.$$

For solids and liquids C_P and C_V are usually quite similar in magnitude but for gases they are significantly different. As $H = U + PV$ (Section 2.7) and, for n moles of perfect gas, $PV = nRT$, we have

$$H = U + nRT.$$

Differentiating we obtain $dH = dU + nRdT$ or $(đq)_P = (đq)_V + nRdT$ (Section 2.7). As $C = (đq/dT)$ (Section 2.2) we may write

$$C_P = C_V + nR$$

For argon at room temperature $C_P = 20.8\,\mathrm{J\,K^{-1}\,mol^{-1}}$, $C_V = 12.5\,\mathrm{J\,K^{-1}}$ $\mathrm{mol^{-1}}$, confirming $C_P - C_V = 8.3\,\mathrm{J\,K^{-1}\,mol^{-1}}$, i.e. R.

Problems

2.1. Joule suggested that water at the bottom of the Niagara Falls, which are 50 m high, should be warmer than that at the top. Estimate the rise in temperature; Joule's own estimate was $\sim 0.10\,\mathrm{K}$. The heat capacity of 1 mol water, 0.018 kg, is $80\,\mathrm{J\,K^{-1}}$. The acceleration due to gravity is $9.8\,\mathrm{ms^{-2}}$.

2.2. An electrical kettle operating at 250 volt and 8 amp holds 1 kg water. If the water is initially at 300 K calculate how long it will take before the water starts to boil. The boiling point of water is 373 K and its heat capacity is $4200\,\mathrm{J\,K^{-1}\,kg^{-1}}$. You can assume the heat capacity does not vary with temperature and that there are no heat losses.

2.3. Calculate the work done against the (standard) atmospheric pressure when a substance expands by $1\,\mathrm{cm^3}$. ($1\,\mathrm{atm} = 10^5\,\mathrm{N\,m^{-2}}$.)

2.4. Calculate the difference between ΔH and ΔU when 1 mol water is boiled at 373 K and 1 atm. The volume of 1 mol of perfect gas at 373 K is $0.03\,\mathrm{m^3}$ and the volume of liquid water may be neglected. ($1\,\mathrm{atm} = 10^5\,\mathrm{N\,m^{-2}}$.)

2.5. A block of metal of mass 1 kg is heated to 400 K and dropped in 0.3 kg of water. The water rises in temperature from 294 K to 300 K. Calculate the heat capaicity of the metal. Take the heat capacity of water as $4200\,\mathrm{J\,K^{-1}\,kg^{-1}}$.

3
Entropy and equilibrium

3.1 Reversibility and equilibrium: a recapitulation

When a chemical reaction proceeds, we have established (by reference to experiment) that energy will be conserved. But we have not found a way of predicting in which direction the reaction will go. In other words we have not found a suitable definition of the position of equilibrium. We have discovered that for molecular systems (which may approach equilibrium by endothermic processes) the energy, unlike the potential energy in mechanical systems, does not provide a sufficient criterion for equilibrium. A new factor must be introduced which will enable us to understand why heat always flows from hot to cold bodies and why a perfect gas will expand to fill its container, even though no loss of energy (by the system) accompanies these processes.

We have observed in our consideration of mechanical systems that if a change occurs so that the system is always in equilibrium, then the change will proceed infinitely slowly and will be capable of doing the maximum amount of work. Such a change we call *reversible*. The conditions that must be satisfied for a reversible change are the same conditions that must be satisfied if the system is to be in a state of equilibrium.

For a reversible change the work done by the system is a maximum. Thus for a reversible change đw is more negative than for the equivalent irreversible spontaneous change.† dU must be the same for a given change whatever way it is carried out. Therefore for a reversible change, since đ$w_{rev} <$ đw_{irr}, we have đ$q_{rev} >$ đq_{irr}. During a reversible change a system absorbs the maximum heat from its surroundings and does the maximum work *on* its surroundings. Observable, spontaneous processes absorb less heat and do less work than the corresponding reversible processes (Figs 3.1 and 3.2).

3.2 Condition of equilibrium

We are now in a position to state the general condition of equilibrium which will apply to both mechanical and molecular systems: *Spontaneous changes are those which, if carried out under the proper conditions, can be made to do work. If carried out reversibly they yield a maximum amount of work. In natural*

---

† Remember, work done *by* the system is negative.

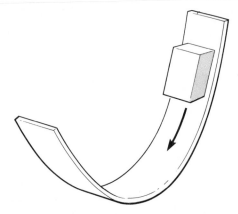

Fig. 3.1. Spontaneous change: occurs at finite speed and requires a finite quantity of work done on the system if it is to be restored to its original state.

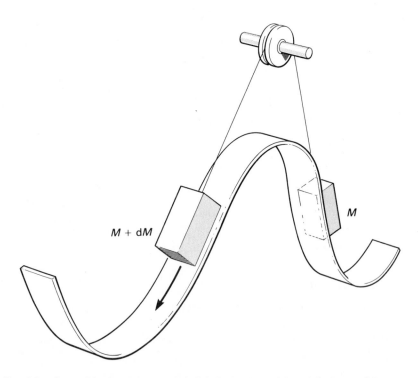

Fig. 3.2. Reversible change: proceeds infinitely slowly, and the original state of the system can be regained by an infinitesimal amount of work.

processes the maximum work is never obtained. This is one of many equivalent statements of the Second Law of Thermodynamics.‡ Like the First Law, it is based on experience; that is, on experimental observations. However, in practice it is sometimes difficult to say just how the work accompanying many spontaneous processes would be obtained. The mixing of two perfect gases is a particularly difficult case.

In a simple mechanical system the capacity to do work is simply the potential energy, and equilibrium is defined by the position of minimum potential energy. However, the total internal energy of a molecular system cannot entirely be transformed into work, and the position of minimum internal energy does not define the equilibrium position in molecular systems. We need a measure of the capacity of such a system to do work and a state function that reflects the loss of the capacity to do work.

3.3 Entropy

In both the examples we have considered earlier (the expansion of a perfect gas into a vacuum, and the flow of heat) the system loses the capacity to do work. This lost capacity is clearly related to $đq$, because $dU = đq + đw$ (see Section 2.5). It is also related to temperature, for if we consider the flow of heat, q, from a hot to a cold reservoir and to a warm reservoir as illustrated in Fig. 3.3 the loss of capacity to do work is clearly greater in the former case. In the latter case work could be obtained from the flow of heat from $T_w \rightarrow T_c$ as this will be a spontaneous process.

We have seen that $đq_{rev}$ is not a convenient measure of 'unavailable' work as it depends on the path; thus q_{rev} is not a state function. A reversible change could, for example, be performed adiabatically or isothermally (see Chapter 8). We have shown also that our measure must involve temperature, because a fixed quantity of heat flowing across a large temperature difference involves a greater loss in the capacity of the system to do work than the same heat flowing across a smaller temperature difference.

It will be shown later that although $q_{rev} = \int đq_{rev}$ is not independent of path, the integral of the heat, $đq_{rev}$, divided by the temperature at which it is transferred—$\int (đq_{rev}/T)$—is independent of the path and depends only on the initial and final states of the system. We define *entropy S*, so that

$$\Delta S = S_B - S_A = \int_A^B \frac{đq_{rev}}{T}$$

Entropy is a state function and is therefore a convenient measure of the loss of capacity of the system to do work. At constant temperature $đq_{rev} = T dS$

‡ Another, and perhaps more familiar, statement of the Second Law is: *heat does not spontaneously flow from a cold to a hot body.*

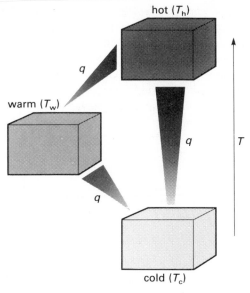

Fig. 3.3. Flow of heat from a hot body to a cold body. The flow may take place either directly or via a warm body.

and as $dU = đq_{rev} + đw_{rev}$ we obtain

$$đw_{rev} = dU - TdS;$$

work = internal energy change − 'unavailable energy'.

   In going from state A to state B, ΔS will always be the same. It will, however, only be equal to $\int (đq_{rev}/T)$ for a reversible path. We have seen that the only condition under which we can consider a change to be reversible is when the system is at equilibrium. Therefore $dS = đq/T$ is a condition of equilibrium.

   For spontaneous changes $đq < đq_{rev}$ (Section 3.1),

and

$$dS > \frac{đq}{T}$$

a condition which holds for observable processes. The equation

$$dS = \frac{đq}{T}$$

is the most general definition of equilibrium available. If we consider an 'isolated' system—one which cannot exchange energy with its surroundings, so that it can neither do work nor absorb heat and for which $đq = 0$—we can obtain as the equilibrium conditions,

$$dS = 0, \qquad S = \text{constant.}$$

For an observable change $dS > 0$ as $dS > đq/T$.

Thus for an isolated system any spontaneous change will tend to produce states of higher entropy until the entropy reaches a maximum value. At this point the system will be in equilibrium and the entropy will remain constant at its maximum value.

3.4 Entropy as a state function

We have said that entropy is a state function but we must justify this statement before proceeding. Traditionally this was done from a consideration of the efficiency of heat engines—a consideration which was a major preoccupation with the pioneers of thermodynamics. As chemists, we shall allow ourselves a short cut by considering only a perfect gas.

From the First Law $dU = đq_{rev} - P dV$ if the system does only PV work. Now for one mole of perfect gas we may write $dU = C_V dT$, as the internal energy of a perfect gas is independent of its volume. Furthermore, as $P = RT/V$,

$$dS = \frac{đq_{rev}}{T} = C_V \frac{dT}{T} + R \frac{dV}{V}.$$

The right-hand side of this equation can be integrated to give

$$S_B - S_A = C_V \ln \frac{T_B}{T_A} + R \ln \frac{V_B}{V_A}.$$

This implies that $đq_{rev}/T$ is an exact differential and that S is a state function for a perfect gas. More general arguments of this type enable us to show that S is a state function for all substances.†

Example

One mole of gaseous oxygen is expanded from $10 \, dm^3$ at 298 K to $20 \, dm^3$ at 400 K. If $C_P = 29.4 \, J \, K^{-1} \, mol^{-1}$ estimate the entropy change associated with the expansion.

For a perfect gas

$$\Delta S = C_V \ln \frac{T_B}{T_A} + R \ln \frac{V_B}{V_A}.$$

---

† This approach was developed by the mathematician Carathéodory (1909).

Since $$C_P = C_V + R \qquad \text{(Section 2.8)}$$
$$C_V = 29.4 - 8.31 = 21.1 \, \text{J K}^{-1} \, \text{mol}^{-1}.$$

Then

$$\Delta S = \left(21.1 \ln \frac{400}{298} + 8.31 \ln \frac{20}{10} \right) \text{J K}^{-1} \, \text{mol}^{-1}$$

$$\underline{\Delta S = 12.0 \, \text{J K}^{-1} \, \text{mol}^{-1}}$$

3.5 Entropy of expansion of a gas

As an example of an entropy change let us consider the isothermal expansion of a perfect gas. From Section 2.5 we know $\Delta U = q + w$. Since for a perfect gas U is independent of volume, $\Delta U = 0$ and $q = -w$: the heat gained from the surroundings is equal to the work done by the system. If, as the gas expands and its pressure drops, the external pressure is continuously adjusted so that $P = P_{ex} + dP$ then the expansion can be carried out *reversibly*, doing the maximum work, and

$$- w_{rev} = \int_A^B P \, dV.$$

For n moles of perfect gas

$$P = nRT/V,$$

$$- w_{rev} = nRT \int_A^B \frac{dV}{V} = nRT \ln \frac{V_B}{V_A},$$

and

$$\Delta S = \frac{q_{rev}}{T} = nR \ln \frac{V_B}{V_A}.$$

If one mole of a perfect gas is expanded from a volume of $0.01 \, \text{m}^3$ to $0.10 \, \text{m}^3$, the entropy change is $\Delta S = 8.3 \ln 10 = 19.1 \, \text{JK}^{-1} \, \text{mol}^{-1}$. This relation will be correct for both reversible and irreversible changes as entropy is a state function and ΔS is independent of the path taken between the states A and B. The heat lost by the surroundings is equal to the heat gained by the gas and thus for a reversible expansion $\Delta S_{overall} = 0$ (as $\Delta S_{overall} = q_{rev}/T - q_{rev}/T$).

However, if we consider the *irreversible* expansion of a perfect gas into a vacuum (doing no work) then $\Delta S_{overall}$ will no longer be zero. For the gas itself $\Delta S = nR \ln V_B/V_A$; however, the changes in the surroundings will be different. The gas does no work and as $q = -w = 0$, the system absorbs no heat from its surroundings, which will be unchanged. Thus

$$\Delta S_{\text{overall}} = nR \ln \frac{V_B}{V_A} + \frac{0}{T} = nR \ln \frac{V_B}{V_A}.$$

As $V_B > V_A$ the total entropy of the system and its surroundings will have increased.

Thus as the entropy change in the gas is the same in both cases we must investigate the surroundings before we can decide whether a change has occurred in a reversible or an irreversible manner. This restricts the usefulness of entropy in defining equilibrium conditions.

3.6 Entropy changes accompanying heat flow

Consider the flow of heat from a large body maintained at temperature T_h to one maintained at T_c. If we imagine the transfer of heat to be reversible at each temperature, the overall entropy change is given by

$$dS = -\frac{dq}{T_h} + \frac{dq}{T_c},$$

or

$$dS = dq\left(\frac{T_h - T_c}{T_h T_c}\right).$$

Thus if the process is an observable one (requiring $T_h > T_c$) then $dS > 0$ and the entropy of the system will increase. Only when $T_h = T_c$ will reversible, equilibrium conditions be obtained.

3.7 Entropy and equilibrium

If we have an *isolated* system which cannot exchange energy with its surroundings, we have seen that entropy will remain constant at equilibrium or will increase if an observable change occurs. Observable changes will continue to occur until the entropy attains a maximum value at which time the system will be in equilibrium. Thus the gas in our example expands until it is uniformly distributed and then no further changes occur. This can be expressed as follows: in a system at constant energy and volume (and which can do no work) the entropy is a maximum at equilibrium: $(dS)_{U,V} = 0$ (see Section 3.3).

This situation may be compared with the criterion for equilibrium in an ordinary mechanical system; at constant entropy and volume (for a system which can do no work) the energy is a minimum (see Section 1.2): $(dU)_{S,V} = 0$.

Though neither of these sets of conditions occurs in the course of normal chemical experiments, they enable us to identify the two factors which drive chemical systems to equilibrium. First there is a tendency to minimize their energy and second there is a tendency to maximize their entropy. In general a compromise between these tendencies must be achieved. The nature of this compromise will be considered later.

3.8 A cosmological aside

We have seen that an *isolated* system will tend to maximize its entropy. Most systems are not isolated and are free to exchange energy with their surroundings. However if we consider both system and surroundings together then we do have an 'isolated system' and $dS > 0$ for any spontaneous process. We may extend our definition of system and surroundings to embrace the whole universe. This leads to a common statement: 'The energy of the universe is constant but the entropy is continually increasing'. The implication, that when the entropy of the universe finally reaches its maximum value no further observable changes will occur, has given much satisfaction to pessimists. When this state occurs all the energy and material of the universe will be *uniformly* distributed and completely unavailable to do work.

3.9 Entropy as a function of pressure and temperature

Pressure. We have already established (Section 3.5) that for n moles of a perfect gas at constant temperature

$$\Delta S = S_B - S_A = nR \ln \frac{V_B}{V_A}.$$

Since $P \propto 1/V$ for a fixed quantity of a perfect gas under isothermal conditions,

$$\Delta S = S_B - S_A = -nR \ln \frac{P_B}{P_A}.$$

If we define S^0 as the entropy of one mole of perfect gas at 1 atm pressure (and at a specified temperature)

$$S = S^0 - R \ln P/P^0 = S^0 - R \ln P/1 \text{ atm.}$$

This equation is often written

$$S = S^0 - R \ln P$$

but P is then the *numerical value* of the pressure expressed in atmospheres. This practice can lead to some confusion over the dimensions of terms in thermodynamic equations and we will adopt the notation $S = S^0 - R \ln (P/\text{atm})$.

The equation tells us that as the pressure is raised the entropy of a gas decreases.

Temperature. As shown (Section 3.3)

$$dS = \frac{dq_{rev}}{T}, \quad \text{and as} \quad C_V = \left(\frac{dq_{rev}}{dT}\right)_V, †$$

and (Section 2.8)

$$C_P = \left(\frac{dq_{rev}}{dT}\right)_P$$

we have $dS = (C_V/T)\,dT$ at constant volume and $dS = (C_P/T)\,dT$ at constant pressure.

Integrating the constant-pressure expression from temperature T_A to temperature T_B, assuming C_P is constant, (as $\int \frac{dT}{T} = \ln T$) we obtain

$$\Delta S = C_P \ln \frac{T_B}{T_A}.$$

If S_0 is the entropy of a substance at the absolute zero of temperature we may write for the entropy at a temperature T, $S(T)$,

$$S(T) = S_0 + \int_0^T \frac{C_P}{T}\,dT.$$

This can be integrated if C_P is known as a function of temperature. The magnitude of S_0 will be the subject of comment later.

Since $dT/T = d(\ln T)$ this equation is sometimes expressed

$$S(T) = S_0 + \int_0^T C_P d(\ln T).$$

Example

The heat capacity of liquid mercury at constant pressure is almost constant at $28\,\text{J K}^{-1}\,\text{mol}^{-1}$ between its freezing point at 234 K and 298 K. Estimate the entropy of liquid mercury at its freezing point. The entropy of liquid mercury at 298 K is $77.4\,\text{J K}^{-1}\,\text{mol}^{-1}$.

If C_P can be regarded as constant the change of entropy with temperature is given by

$$\Delta S = C_P \ln \frac{T_B}{T_A}.$$

† $C_V = \left(\frac{dq}{dT}\right)_V$ and $C_P = \left(\frac{dq}{dT}\right)_P$ for all transfers of heat whether reversible or irreversible.

For the data given

$$\Delta S = 28 \ln\frac{298}{234} = 6.8 \, \text{J K}^{-1}\,\text{mol}^{-1}$$

$$S_{234} = S_{298} - 6.8 = 70.6 \, \text{J K}^{-1}\,\text{mol}^{-1}.$$

3.10 Molecular basis of entropy

It should be clear that entropy is related to the *uniformity* of systems. As gradients in temperature or concentration are eliminated the entropy increases. This idea can be placed on a quantitative basis if we consider a specific case; the expansion of a perfect gas. This discussion involves a consideration of the molecular nature of matter. It is not essential to the development of classical thermodynamics, but provides valuable insight.

Consider M gas molecules first contained in one half of a vessel, as illustrated in Fig. 3.4, and then allowed to occupy the whole volume. The probability that state A could occur by chance is $(\frac{1}{2})^M$, that is the same as the chance that M objects will all be in one of two boxes between which they have been randomly distributed. We can write the relative probability of state A with respect to state B as

$$\frac{p_A}{p_B} = (\tfrac{1}{2})^M.$$

If instead of making $V_A/V_B = \frac{1}{2}$ we had selected arbitrary values we can show:

$$\frac{p_A}{p_B} = \left(\frac{V_A}{V_B}\right)^M \quad \text{i.e.} \quad \ln\frac{p_B}{p_A} = M \ln\frac{V_B}{V_A}.$$

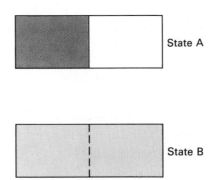

State A

State B

Fig. 3.4. Expansion of a gas. In State A the gas occupies only half the vessel. In State B it is uniformly distributed throughout the whole vessel.

In going from state A to state B we have gone from a state of low probability to one of high probability. As seen in Section 3.5

$$S_B - S_A = R \ln(V_B/V_A)$$

for one mole of gas. We can see that if M is numerically equal to the Avogadro's constant, N_A, then the entropy change per mole is

$$S_B - S_A = \frac{R}{N_A}(\ln p_B - \ln p_A) = \frac{R}{N_A}\ln\frac{p_B}{p_A}.$$

Thus the entropy of a system in any particular state is proportional to ln p where p is the 'probability' of the system. We can write, following Boltzmann,

$$S = k \ln W$$

where k is Boltzmann's constant (R/N_A) and W the number of 'microstates' or 'complexions' of the system. This is a difficult concept. It is the number of ways the state can be made up by specifying the positions and velocities of the atoms that comprise it. Feynman expressed it thus: W is 'the number of ways the inside of a system can be made up if the outside stays the same'. For systems with about 10^{23} molecules W can be very large, of the order of $10^{10^{23}}$.

An important branch of physical chemistry called statistical mechanics is concerned with the calculation of W, and hence the evaluation of the thermodynamic properties of molecular systems without recourse to thermo-dynamic experiment. An introduction to this subject is given in Chapter 9.

For a simple mechanical system the entropy difference between one state and another is usually negligible. Let us consider an object that can be thrown from a distance and so be randomly distributed between two boxes, one twice the size of the other. The probability of the object landing in the larger box (B) will be twice that of it landing in the smaller box (A).

$$\Delta S = S_B - S_A = k \ln 2$$
$$\approx 1.38 \times 10^{-23} \times 2.3 \times 0.301 \text{ J K}^{-1};$$

i.e.

$$\Delta S \approx 1 \times 10^{-23} \text{ J K}^{-1}.$$

This is a very small entropy difference which may be neglected without introducing any significant inaccuracy when performing calculations on such a system. This is why only energy need be considered when determining the position of equilibrium in mechanical systems.

However, if we consider two states for a mole of gas, one in which the gas is contained in a volume twice as large as the other, we obtain (as calculated earlier in the chapter)

$$\Delta S = S_B - S_A = k \ln 2^{N_A} = R \ln 2 = 8.3 \times 2.3 \times 0.301 = 5.8 \text{ J K}^{-1}$$

a substantial and far from negligible contribution.

3.11 Statistical basis of the Second Law

When we consider entropy from a molecular viewpoint we realize that it is not *impossible* for all the molecules in a container to be in one half at the same time—only *very improbable*. It is very improbable that all the air molecules in the room in which one is sitting will congregate in one corner leaving one with no air to breathe, but it is not strictly speaking impossible. To see just how improbable let us consider our mole of gas contained in two volumes, one half the size of the other. For the process B → A (see Fig. 3.4)

$$\Delta S = -5{\cdot}8 \text{ J K}^{-1} = \frac{R}{L} \ln \frac{p_A}{p_B}.$$

$$\frac{p_A}{p_B} = \exp\left(-\frac{5.8 \times 6 \times 10^{23}}{8.3}\right) \approx \exp(-10^{23}).$$

In other words it is very, very unlikely that a mole of gas should find itself in one half of its container. (It is even more unlikely that such a disturbing occurrence should occur in one's room—as it will contain many moles of gas.) As Boltzmann[†] put it, when discussing the probability of such unlikely events occurring in the future, 'by this time there will have been many years in which every inhabitant of a large country will have committed suicide on the same day as every building has burnt down. . . . Insurance companies get away without worrying about this.' In just the same way we, in our capacity as scientists (or indeed as air breathers) can ignore such remote possibilities. The consequences of small fluctuations in the distribution of molecules and their energies can, however, sometimes be of importance. The Brownian motion of bacteria suspended in a liquid arises from the fluctuating pressures on the particles due to molecular bombardment.

3.12 Magnitudes of entropy changes

The entropy of a chemical system is to a large degree determined by the 'freedom' possessed by the molecules in the system. In solids where the molecules or atoms are tightly bound the entropy is low. In gases where the molecules are free to move in a large volume the entropy is high. Liquids are intermediate in their properties.

When a liquid vaporizes the molecules go from a state of modest freedom to one of high freedom. The entropy change associated with vaporization is therefore positive:

$$\Delta S_{vap} = \frac{q_{rev}}{T} = \frac{\Delta H_{vap}}{T}.$$

[†] L. Boltzmann (1964) *Lectures on Gas Theory* (trans. S. G. Brush). University of California Press, Berkeley, California. Reprinted by permission of the Regents of the University of California.

Table 3.1 Molar entropies of substances at 298 K and 1 atm pressure

Substance	$S^\circ / J K^{-1} mol^{-1}$
Diamond (solid)	2.4
Silver (solid)	42.7
Water (liquid)	69.9
Argon (gas)	154.7
Carbon dioxide (gas)	213.6

For benzene $\Delta H_{vap} = 30.7$ kJ mol^{-1} and the normal boiling point is 353.3 K. Thus $\Delta S_{vap} = 87.0$ J K^{-1} mol^{-1}. Indeed it is found that for most non-polar liquids the entropy of vaporization at their normal boiling point is approximately 90 J K^{-1} mol^{-1}. This generalization is called Trouton's Rule.

The entropy change associated with a chemical reaction depends on the nature of the reactants and products. The reaction

$$CaCO_3(s) \rightleftharpoons CaO(s) + CO_2(g)$$

has a large and positive entropy change of 160 J K^{-1} mol^{-1} as the solids have low entropy, whereas the CO_2 as a gas has high entropy. Even in a reaction involving two gases significant entropy changes may occur. Thus for $N_2O_4 \rightarrow 2NO_2$, $\Delta S = 177$ J K^{-1} mol^{-1} with all gases at 1 atm pressure. The entropy of the NO_2 is greater when it is present as independent molecules than when it is bound forming N_2O_4 molecules. This positive entropy change favours the dissociation of the dimer which is 20 per cent dissociated at room temperature despite its relatively strong binding energy which would, in the absence of entropy considerations, suggest that little dissociation could be expected at such a temperature.

3.13 Heat engines

As mentioned earlier the pioneers of modern thermodynamics reached the concept of entropy from an analysis of the efficiency of heat engines.† A heat engine is a device for converting heat, usually generated by combustion, into work. The steam engine and the internal combustion engine are the most familiar examples. We can represent a heat engine schematically as shown in Fig. 3.5. Each cycle takes heat (q_h) from a high-temperature reservoir, uses some of it to generate work, w, as by the expansion of a gas against a piston, and rejects unused heat (q_c) to a colder reservoir. The maximum work we can obtain from such an operation is that generated when all the processes are

---

† Such an analysis on the basis of the so-called Carnot cycle is available in most standard physical chemistry text books.

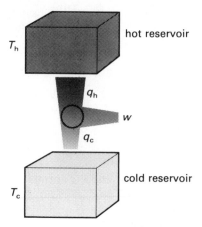

Fig. 3.5. Schematic diagram of a heat engine.

reversible. Then the total entropy of the system must remain constant.

$$\Delta S_h + \Delta S_c = 0, \qquad - q_h/T_h + q_c/T_c = 0,$$

and $q_h/q_c = T_h/T_c$.

But, by the Principle of Conservation of Energy $w = q_h - q_c$ and

$$\frac{w}{q_h} = \frac{q_h - q_c}{q_h},$$

$$\frac{w}{q_h} = 1 - \frac{q_c}{q_h} = 1 - \frac{T_c}{T_h},$$

$$\frac{w}{q_h} = \frac{T_h - T_c}{T_h}.$$

This factor, the ratio of work obtained to the total heat taken from the high-temperature source, is called the thermodynamic efficiency of the engine. For a steam engine using steam at 400 K and rejecting it from the engine at 300 K the thermodynamic efficiency would be

$$\frac{400 - 300}{400} = 25 \text{ per cent.}$$

Only one-quarter of the heat is converted into work—thus the heat engine is a wasteful way of producing work (from a thermodynamic standpoint if not an economic one). We see that when $T_h = T_c$ we can get no work—an isothermal cycle cannot use heat to give work. If $T_c = 0$ then 100 per cent efficiency can be attained. This is in accord with the fact that heat is associated with molecular motion. The molecules at absolute zero would be in a state such that their motion could not be further diminished and all their original thermal energy would have been converted into work.

It is worth noting that if we reverse the operation of the heat engine (that is, use work to take heat from a cold reservoir and to release it at higher temperatures) we get thermodynamic efficiencies greater than 100 per cent, which at first sight may appear paradoxical. As

$$\frac{q_h}{w} = \frac{T_h}{T_h - T_c},$$

if $T_h = 300$ and $T_c = 280$ the thermodynamic efficiency is 1500 per cent! Many large buildings are heated by such 'heat pumps' with a typical operating efficiency of about 400 per cent; that is, four times as much heat is produced as work is put in, whereas an electric fire gives only about 100 per cent. The reasons for the limited uses of heat pumps are economic—they are costly to install.†

Problems

3.1. Calculate the entropy change if $0 \cdot 011$ m^3 of a perfect gas at 273 K and 1 atm pressure is compressed to 10 atm pressure.

3.2. The heat capacity of gaseous argon at constant pressure is 20.8 J K^{-1} mol^{-1}. Estimate the entropy change when one mole of argon is heated from 300 K to 1200 K at 1 atm pressure.

3.3. Estimate the entropy change when one mole of water is vaporized at 373 K. The enthalpy change on vaporization is 40.7 kJ mol^{-1} at this temperature.

3.4. Calculate the thermodynamic efficiency of a heat engine operating between the temperatures 600 K and 400 K.

3.5. Calculate the entropy change when one mole of ice at 268 K is melted to form water at 323 K. The heat capacity of ice is 3.8 J K^{-1} mol^{-1}, that of water is 75 J K^{-1} mol^{-1} and the enthalpy of fusion of ice at 273 K is 6.02 kJ mol^{-1}.

3.6. Calculate the entropy change when one mole of cadium vapour at 1 atm pressure is heated from 1040 K to 1100 K and subsequently compressed to a pressure of 6 atm. You may assume that the vapour follows perfect gas behaviour.

$C_v[Cd(g)] = 12.5$ J K^{-1} mol^{-1}.

---

† For a popular account of the applications of heat pumps see J. F. Sandfoot, *Scientific American*, **184**, May 1951, p. 54.

4
Equilibrium in chemical systems

4.1 Free energy

We have seen that for systems at constant energy the position of equilibrium can be defined by the condition of maximum entropy. Since a system and its surroundings taken together could be regarded as a new system whose energy is constant, the position which leads to the maximum entropy for system *and* surroundings is the equilibrium position. However, it is more convenient to have a definition of the position of equilibrium which can be expressed in terms of the properties of the system alone and which does not require a knowledge of changes taking place in the surroundings. To find such a condition we return to the concept of maximum work. In Section 3.2 we defined the position of equilibrium in terms of the capacity of a system to do work. When during a change a system does the maximum possible work, that change is reversible and the system is, at every stage, at equilibrium. We shall now introduce thermodynamic functions which are related to the maximum amount of work that can be obtained from a system at constant temperature. These provide a most useful definition of the position of equilibrium in chemical and physical systems. For reversible conditions, as $dU = đq_{rev} + đw_{rev}$ (Section 2.5) and $dS = đq_{rev}/T$ (Section 3.3), we have

$$dU = T\,dS + đw_{rev}.$$

and

$$đw_{rev} = dU - T\,dS.$$

If we define a state function, the *Helmholtz free energy A*, such that

$$A = U - TS$$

then at constant temperature

$$dA = dU - T\,dS$$

and

$$đw_{rev} = dU - T\,dS = dA.$$

This is a condition that must be satisfied by a reversible process and hence $đw = dA$ is also a condition for the system to be at equilibrium. If during a reversible change the system does work, $đw_{rev}$ will be negative. dA will also be negative and A will decrease. A is the equivalent function, in this molecular system at constant T and V, to the energy U in a mechanical system; *it is a measure of the maximum amount of work the system can do on its surroundings.*

For a spontaneous process, the system may do work but the work will be less than for the equivalent reversible change: thus đw will be less negative than đw_{rev}.

$$đw > đw_{rev}$$

and

$$đw > dA.$$

For the spontaneous process both đw and dA will be negative but đw will be less negative than dA. The work done *by* the system ($- đw$) will be less than the decrease in A, thus only part of the system's change in free energy will be obtained as work.

If a system is not harnessed to do work đ$w = 0$ and, as đ$w_{rev} < đw$, d$A < 0$, and dA for a spontaneous process will again be negative. Eventually if the spontaneous change continues, A will reach a minimum value and no further work can be obtained from the system even under reversible conditions.

Then

$$đw_{rev} = 0$$

and

$$dA = 0.$$

Thus the condition of equilibrium in a system at constant temperature and volume which does no work is

$$dA = 0.$$

Spontaneous processes may occur in such a system when it is not at equilibrium with a consequent decrease in the free energy. When A is a minimum and d$A = 0$ no further spontaneous changes can occur and the system is at equilibrium. Again we see the parallel between the free energy A and the potential energy in a mechanical system. If the latter system is not harnessed to do work the position of equilibrium could be defined in terms of minimum energy.

4.2 Gibbs free energy

As chemists, we are often interested in systems at constant pressure and temperature rather than at constant volume. Under constant pressure conditions we can write

$$đw_{rev} = - P\,dV + đw_{additional},$$

where đ$w_{additional}$ is the work other than PV work done on the system.

Electrical work done on a solution being electrolysed would be an example of this additional work. At equilibrium

$$đw_{rev} = dU - đq_{rev} \qquad \text{(Section 2.5)}$$

and

$$đq_{rev} = T\,dS \qquad \text{(Section 3.3)},$$

thus

$$đw_{additional} - P\,dV = dU - T\,dS.$$

We may define a further state function, the *Gibbs free energy G*, such that

$$G = U + PV - TS = H - TS.$$

At constant P and T

$$dG = dU + P\,dV - T\,dS.$$

As $đw_{additional} = dU + P\,dV - T\,dS$ for a reversible change we have

$$đw_{additional} = dG.$$

This is also the condition for equilibrium in a system at constant temperature and pressure. For a spontaneous process $đw_{additional}$ will be less negative than for the corresponding reversible change and not all the decrease in G will be obtained as additional work.† If a system at constant T and P does no additional work then the condition for equilibrium is $dG = 0$. G will be at a minimum when the system is at equilibrium. In a molecular system at constant T and P, G is the measure of the maximum work (other than PV work) which may be obtained from the system. When the system is capable of doing no work and is at equilibrium, $dG = 0$ and G is a minimum. Again we can point to the analogy between free energy in molecular systems and energy in mechanical systems (Fig. 4.1).

G is a state function and therefore ΔG has a definite value for any change. However it is equal to the maximum available work only for a change that is carried out at constant T and P.

In the equation

$$\Delta G = \Delta H - T\Delta S$$

we have found the proper balance at constant pressure and temperature between the tendencies of a system to maximize its entropy and to minimize its energy (or, at constant pressure, more strictly its enthalpy). At higher temperatures the contribution of entropy change to the free-energy change,

---

† As $P\,dV$ is the same for a reversible or an irreversible change at constant pressure, the additional work will behave in the same way as the total work. Thus the system will do the maximum additional work in a reversible change.

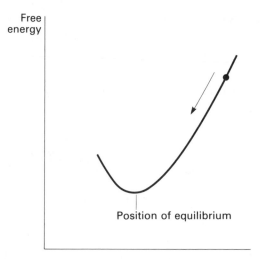

Fig. 4.1. Position of equilibrium in terms of Gibbs free energy for a system at constant pressure and temperature.

$-T\Delta S$, becomes relatively more important. As we have seen, for the reaction

$$N_2O_4 \rightleftharpoons 2NO_2$$

ΔH is positive, since there is an energy that holds together the dimer, and ΔS is positive because the separated monomers have more freedom to move than when bound in dimers. At low temperatures ΔG is positive and little dissociation will occur. At high temperatures the favourable entropy term $(-T\Delta S < 0)$ will dominate and increased dissociation will be observed.

Having found how the position of equilibrium can be defined at constant temperature and pressure we shall now restrict ourselves to these conditions. We shall investigate the properties of the Gibbs free energy G which provides our criterion of equilibrium.

As

$$G = U + PV - TS \qquad \text{(Section 4.2)},$$

$$dG = dU + P\,dV + V\,dP - S\,dT - T\,dS.$$

But for a system that does only PV work,

$$\text{as} \quad dU = \text{d}q_{rev} + \text{d}w_{rev} \qquad \text{(Section 2.5)},$$

$$dU = T\,dS - P\,dV,$$

and

$$dG = V\,dP - S\,dT\,.$$

This is a most important thermodynamic equation for chemists as it tells us how free energy, and hence equilibrium position, varies with pressure and temperature.

4.3 Pressure-dependence of free energy

At constant temperature $dT = 0$ and

$$dG = V dP, \quad \text{or} \quad \left(\frac{\partial G}{\partial P}\right)_T = V.$$

For n moles of a perfect gas $PV = nRT$ (Section 1.6), so

$$dG = nRT \frac{dP}{P}.$$

For a change in pressure from P_A to P_B.

$$\Delta G = G_B - G_A = nRT \int_{P_A}^{P_B} \frac{dP}{P},$$

$$\Delta G = nRT \ln \frac{P_B}{P_A}. \dagger$$

We usually relate the free energy of a gas to the *standard free energy* G^0. This is defined as the free energy of one mole of the gas at one atmosphere pressure.
   Then

$$G = G^0 + RT \ln P/P^0.$$

As P^0 has the value 1 atmosphere,

$$G^0 = G^0 + RT \ln P/1 \text{ atm}.$$

We write

$$G = G^0 + RT \ln (P/\text{atm}).$$

This equation is sometimes expressed

$$G = G^0 + RT \ln P$$

---

† We could have obtained this equation from $\Delta G = \Delta H - T\Delta S$ as we have already established (Section 3.9) that $\Delta S = -nR \ln P_B/P_A$ and that ΔH is independent of pressure for a perfect gas (so that at constant temperature $\Delta H = 0$).

where P now represents a *dimensionless* ratio. This can lead to confusion when equations of this type are used.

Example

The standard free energy of nitrogen is defined as zero at 298 K and 1 atm pressure. Calculate its value at 10 atm and 0.20 atm at the same temperature.

For a perfect gas

At 10 atm

$$G = G^0 + RT \ln (P/\text{atm})$$
$$= 0 + RT \ln 10$$
$$G = 2476 \text{ J mol}^{-1}.$$

At 0.2 atm

$$G = 0 + 8.31 \times 298 \ln (0.2)$$
$$G = -1238 \text{ J mol}^{-1}.$$

4.4 Temperature variation of free energy

Recalling the basic equation $dG = V dP - S dT$, at constant pressure $dP = 0$, therefore

$$dG = -S dT, \quad \text{and} \quad \left(\frac{\partial G}{\partial T}\right)_P = -S.$$

But

$$G = H - TS;$$

therefore

$$G = H + T\left(\frac{\partial G}{\partial T}\right)_P.$$

If we divide throughout by T^2 we obtain

$$-\frac{G}{T^2} + \frac{1}{T}\left(\frac{\partial G}{\partial T}\right)_P = -\frac{H}{T^2}.$$

Since

$$-\frac{G}{T^2} + \frac{1}{T}\left(\frac{\partial G}{\partial T}\right)_P = \left[\frac{\partial \left(\frac{G}{T}\right)}{\partial T}\right]_P,$$

$$\left[\frac{\partial\left(\frac{G}{T}\right)}{\partial T}\right]_P = -\frac{H}{T^2},$$

and

$$\left[\frac{\partial\left(\frac{\Delta G}{T}\right)}{\partial T}\right]_P = -\frac{\Delta H}{T^2}.$$

These are the *Gibbs–Helmholtz equations*. They are very important because they relate the temperature dependence of free energy, and hence the position of equilibrium, to the enthalpy change. We shall illustrate their application in the next section.

4.5 Phase equilibria

Equilibrium between the states of matter can be understood in terms of the equation $G = H - TS$ (Section 4.2). The phase with the lowest free energy under any conditions is the most stable. For solids H is relatively large and negative† because of the strong binding forces in solids but S is small as the molecules have little freedom, so at low temperatures solids are the most stable phase.

For gases H is close to zero as there are no strong interactions between the molecules, but S is large because the molecules have a large amount of room in which to move. Thus gases are the stable phase at high temperatures. There is an intermediate region where liquids, which have a less negative H but larger S than solids, are stable. This is illustrated in Fig. 4.2 where the free energy is plotted as a function of temperature. The slopes of the lines are determined by the entropy of the phases because $\left(\frac{\partial G}{\partial T}\right)_P = -S$ (Section 4.4). The phase with the highest entropy, the gas phase, has the largest negative slope and its free energy is the lowest at high temperatures. Where the lines intersect, the free energies of the two phases represented by the lines are equal. As $\Delta G = 0$, and $\Delta G = \Delta H - T\Delta S$ we obtain for the transition between two phases in equilibrium $\Delta H = T\Delta S$. Thus

$$\Delta S_{fus} = \frac{\Delta H_{fus}}{T_{fus}} \quad \text{and} \quad \Delta S_{vap} = \frac{\Delta H_{vap}}{T_{vap}}.$$

† Relative to the enthalpy of the substance in the perfect-gas state and under the same conditions.

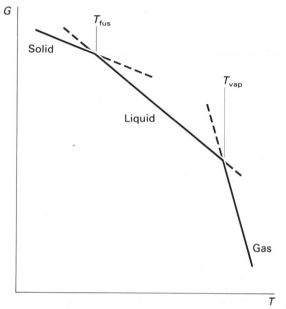

Fig. 4.2. Gibbs free energy as a function of temperature for a pure substance.

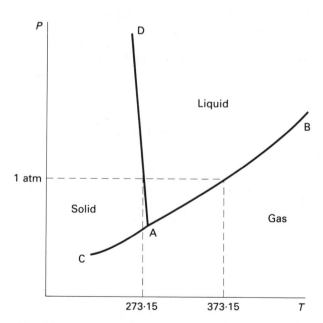

Fig. 4.3. Phase diagram for water (schematic: not to scale).

The information from such diagrams, if available at a series of pressures, may be represented in *phase diagrams* such as Fig. 4.3. Line AB is the vapour pressure curve for water. AD is the melting curve. AC gives the vapour pressure of ice. Point A, at which all three lines meet, is called the *triple point* (for water it is at 273.16 K and 4.58 mmHg pressure, i.e. 10.61 kPa).

Whereas the temperature-dependence of the free energy of a phase is related to its entropy, the pressure dependence is related to its volume. As shown in Section 4.3, $(\partial G/\partial P)_T = V$, and the variation in the equilibrium position between phases (as represented by the lines in Fig. 4.3) depends on the volume change associated with the phase transition. Thus as ΔV for melting is small compared with the volume change associated with vaporization, the melting point is very much less sensitive to pressure than the boiling point.

4.6 Clapeyron equation

We can now apply the thermodynamic considerations of the previous section in a quantitative manner to obtain an important relation. Consider two phases, a liquid and its vapour, in equilibrium at temperature T and pressure P. If we alter the conditions to $T + dT$ and $P + dP$ then, as $dG = VdP - SdT$ (Section 4.2), we obtain for the liquid $dG(l) = V(l)dP - S(l)dT$, and for the vapour $dG(g) = V(g)dP - S(g)dT$. If under the new conditions the phases are still in equilibrium then $dG(l) = dG(g)$.

Equating the free-energy changes,

$$V(l)dP - S(l)dT = V(g)dP - S(g)dT,$$

and

$$\frac{dP}{dT} = \frac{S(g) - S(l)}{V(g) - V(l)} = \frac{\Delta S_{vap}}{\Delta V_{vap}}.$$

As the two phases are in equilibrium,

$$\Delta G_{vap} = \Delta H_{vap} - T\Delta S_{vap} \doteq 0,$$

$$\Delta S_{vap} = \frac{\Delta H_{vap}}{T};$$

and

$$\frac{dP}{dT} = \frac{\Delta H_{vap}}{T\Delta V_{vap}}$$

where T is the boiling point at the pressure under consideration. The equation

$$\left(\frac{dP}{dT}\right)_{equil} = \frac{\Delta H}{T\Delta V}$$

is known as the Clapeyron equation. It is exact and applies to equilibrium between any two phases, that is the melting process as well as vaporization.

Example

At 273.16 K the enthalpy change on fusion of water is 6.0 kJ mol^{-1} and the corresponding volume change -1.6×10^{-6} m^3 mol^{-1}. Estimate the temperature at which ice will melt at 1000 atm pressure (take 1 atm $= 10^5$ N m^{-2}).

We can write the Clapeyron equation

$$\frac{\Delta T}{\Delta P} = T\frac{\Delta V}{\Delta H}$$

and thus

$$\Delta T = \frac{273 \times (-1.6 \times 10^{-6}) + 10^3 \times 10^5}{6 \times 10^3} \, K$$

$$= -7.3 \, K.$$

Thus we estimate that ice will melt at $(273.16 - 7.3)$ K $= 265.9$ K.

4.7 Clausius–Clapeyron equation

When applied to vaporization the Clapeyron equation can be modified to give another useful and important relation. We use the equation derived in Section 4.6,

$$\frac{dP}{dT} = \frac{\Delta H_{vap}}{T\Delta V_{vap}},$$

together with

$$\Delta V_{vap} = V(g) - V(l).$$

At room temperature and 1 atm pressure $V(g) \approx 24\,000$ cm^3 and $V(l) \approx 100$ cm^3 for one mole of substance, so that $V(g) \gg V(l)$ and we can replace ΔV_{vap} by $V(g)$. Again, if the vapour follows the perfect-gas equation, for one mole we have

$$V(g) = \frac{RT}{P} \qquad \text{(Section 1.5),}$$

and

$$\frac{dP}{dT} = \frac{\Delta H_{vap}}{RT^2} P.$$

Thus

$$\frac{d \ln P}{dT} = \frac{\Delta H_{vap}}{RT^2}.$$

If ΔH is independent of temperature then

$$\ln P = -\frac{\Delta H_{vap}}{RT} + \text{const.}$$

These equations, which relate the temperature dependence of the vapour pressure of a liquid to ΔH_{vap}, its enthalpy change per mole on vaporization, are called the Clausius–Clapeyron equations. Unlike the Clapeyron equation, they are not exact, as a number of approximations were introduced in their derivation, but they are nevertheless extremely valuable.

4.8 The vapour pressure of liquids

Let us now consider the vaporization of a liquid from the standpoint of the free-energy change on vaporization. The free energy of one mole of perfect vapour is given by

$$G(g) = G^0(g) + RT \ln (P/\text{atm}) \qquad \text{(Section 4.3)},$$

where P is the vapour pressure of the liquid. $G^0(g)$ is the free energy of one mole of vapour at 1 atm pressure. The free energy of one mole of liquid will be simply $G^0(l)$ as we can assume that the free energy of a condensed phase is virtually independent of pressure. The change in free energy when vaporization of one mole of liquid occurs producing one mole of vapour at its equilibrium pressure, P, is

$$\Delta G = G(g) - G(l) = G^0(g) - G^0(l) + RT \ln (P/\text{atm}).$$

As the liquid and its vapour are in equilibrium there is no change in free energy when a quantity of the liquid is vaporized. Thus $G(g) = G(l)$ and

$$\Delta G^0_{vap} = -RT \ln (P/\text{atm}).$$

This equation tells us that the vapour pressure of a liquid is determined by the free energy change when one mole of liquid is vaporized to produce one mole of vapour *at one atmosphere pressure*. At the normal boiling point where the liquid is in *equilibrium* with its vapour at one atmosphere pressure, $P = 1$ and $G^0_{vap} = 0$.

To find the temperature variation of vapour pressure we use the Gibbs–Helmholtz equations derived earlier (Section 4.4).

$$\left[\frac{\partial \left(\frac{\Delta G}{T} \right)}{\partial T} \right]_P = - \frac{\Delta H}{T^2}$$

$$\frac{d \ln P}{dT} = - \frac{1}{R} \left[\frac{\partial \left(\frac{\Delta G^0_{vap}}{T} \right)}{\partial T} \right] = \frac{\Delta H^0_{vap}}{RT^2} ;\dagger$$

$$\frac{d \ln P}{dT} = \frac{\Delta H^0_{vap}}{RT^2} \cdot \ddagger$$

This is the Clausius–Clapeyron equation we derived in another way in the previous section.

The vapour pressure of n-butane has been measured, with the results given in Table 4.1.

We plot lg P against $1/T$ as in Fig. 4.4, and if ΔH^0_{vap} is independent of temperature over the range of our data, we have lg $P = - \Delta H^0_{vap}/2.3\,RT$ + const. The slope of the plot is $- \Delta H^0_{vap}/2.3\,R$. The slope is found to be $- 1.3 \times 10^3$ therefore $\Delta H_{vap} = 24.8$ kJ mol^{-1} (an average value in the temperature range 200–270 K). Closer inspection shows the plot to be slightly curved, and more careful analysis of this data gives ΔH^0_{vap} at the normal boiling point as 22.38 kJ mol^{-1}.

Table 4.1 The vapour pressure of n-butane

Experimental results†		Our calculation	
T/K	P/mmHg	10^3 K/T	lg(P/mmHg)‡
195.12	9.90	5.125	0.9956
212.68	36.26	4.702	1.5595
226.29	85.59	4.419	1.9324
262.28	503.34	3.812	2.7019
272.82	764.50	3.665	2.8834

† Aston, J. G. and Messerly, G. H. (1940). *J. Am. Chem. Soc.*, **62**, 1917.
‡ We use the convention that lg $x = \log_{10} x$, just as ln $x = \log_e x$.

---

† The Gibbs–Helmholtz equation is strictly applicable only at constant pressure, but as the vapour pressure of a liquid is relatively insensitive to changes in the applied pressure we need not trouble with this restriction.
‡ Note: As $d \ln P = dP/P$ it is a dimensionless quantity.

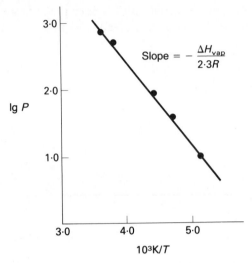

Fig. 4.4. Logarithm of the vapour pressure of liquid n-butane as a function of reciprocal temperature.

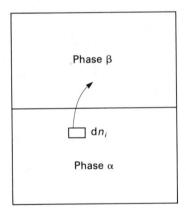

Fig. 4.5. A component in equilibrium between two phases α and β.

4.9 Chemical potential

In our discussion so far we have restricted ourselves to systems containing only one chemical component. We must now consider how we can treat many-component systems, in particular those systems in which the chemical composition changes, as in systems in which chemical reactions take place.

For a pure substance or a system of constant chemical composition

$$dG = VdP - SdT \quad \text{(Section 4.2)}.$$

If the numbers of moles of various components of the system, $n_1, \ldots n_i$, vary we may add further terms to this equation:

$$dG = V dP - S dT + \left(\frac{\partial G}{\partial n_1}\right)_{T,P,n_j} dn_1 + \cdots + \left(\frac{\partial G}{\partial n_i}\right)_{T,P,n_j} dn_i,$$

where the subscript n_j indicates that the quantities of all components, except the one in the derivative, are kept constant. We may define the *chemical potential* μ_i of the ith component

$$\mu_i = \left(\frac{\partial G}{\partial n_i}\right)_{T,P,n_j};$$

then

$$dG = V dP - S dT + \sum_i \mu_i dn_i.$$

This is sometimes called the fundamental equation of chemical thermodynamics. μ_i may be thought of as the increase in the free energy of the system when one mole of component i is added to an infinitely large quantity of the mixture so that it does not significantly change the overall composition. Chemical potential is an intensive property and can be regarded as providing the force which drives chemical systems to equilibrium. Consider a chemical i distributed between two phases α and β as illustrated in Fig. 4.5. Let its chemical potential be $\mu_i(\alpha)$ and $\mu_i(\beta)$ in these phases. At constant T and P if we transfer dn_i moles of i from α to β,

$$dG = [\mu_i(\beta) - \mu_i(\alpha)] dn_i.$$

At equilibrium $dG = 0$ and as we can always consider a small but non-zero value of dn_i this means that $\mu_i(\alpha) = \mu_i(\beta)$ is the condition for equilibrium. Thus for a system at constant pressure and temperature the chemical potential of each component must be equal in all parts of the system. Thus

$$\mu_i(\alpha) = \mu_i(\beta)$$

$$\mu_j(\alpha) = \mu_j(\beta) \quad \text{etc.}$$

This is a most useful definition of the position of chemical equilibrium.

4.10 Chemical potential and free energy

For a pure substance the chemical potential $\left(\frac{\partial G}{\partial n_i}\right)_{T,P,n_j}$ is simply the molar free energy G/n_i (Fig. 4.6). Thus for one mole of gas

$$G = G^0 + RT \ln (P/\text{atm}) \qquad \text{(Section 4.3)}$$

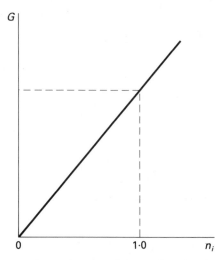

Fig. 4.6. Gibbs free energy as a function of the number of moles of a pure substance.

and $\mu = \mu^0 + RT \ln (P/\text{atm})$. Mixtures of perfect gases behave as if each gas were alone in the container. The thermodynamic properties of the gases making up such a mixture can be expressed in terms of their partial pressures, the pressures they would generate if alone in the container (Section 1.5). Thus

$$\mu_i = \mu_i^0 + RT \ln (P_i/\text{atm})$$

where P_i represents the numerical value of the partial pressure of component i when this is expressed in atmospheres.

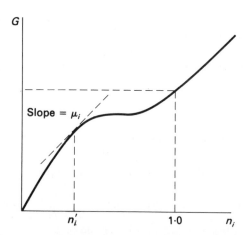

Fig. 4.7. Gibbs free energy as a function of the number of moles of substance i added to a mixture. The slope is the chemical potential of i, μ_i.

The chemical potential of a hypothetical mixture is illustrated in Fig. 4.7. At n_i', $\mu_i = \left(\dfrac{\partial G}{\partial n_i}\right)$ is clearly not equal to G/n_i in a system that behaves like the one illustrated. μ_i is in fact the increase in free energy that occurs when a mole of i is added to an infinitely large quantity of the mixture so that the composition does not change. Differential quantities of this type are called *partial molar quantities*. Thus

$$\mu_i = \left(\frac{\partial G}{\partial n_i}\right)_{T,P,n_j} = \bar{G}_i$$

where $\bar{G}_i$ is the partial molar free energy of component i in the system. The subscript n_j indicates that the amounts of species other than i are constant. We can define other partial molar quantities:

$$\left(\frac{\partial V}{\partial n_i}\right)_{T,P,n_j} = \bar{V}_i$$

$$\left(\frac{\partial S}{\partial n_i}\right)_{T,P,n_j} = \bar{S}_i.$$

$\bar{V}_i$ and $\bar{S}_i$ are the partial molar volume and entropy. As

$$\left(\frac{\partial G}{\partial P}\right)_T = V, \qquad \left(\frac{\partial \mu_i}{\partial P}\right)_T = \frac{\partial}{\partial n_i}\left(\frac{\partial G}{\partial P}\right)_T = \bar{V}_i.$$

A similar argument gives

$$\left(\frac{\partial \mu_i}{\partial T}\right)_P = -\bar{S}_i.$$

Partial molar quantities play an important role in the study of non-ideal mixtures but we have to use them to only a limited extent in elementary thermodynamics. They can usually be replaced by the corresponding molar quantities. Thus, in simple calculations involving perfect gases or ideal solutions, $\bar{V}_i$ can be replaced by the volume of one mole of pure i in the appropriate physical state.

4.11 Equilibrium between gaseous reactants

Consider the equilibrium

$$A(g) \rightleftharpoons B(g).$$

This is the simplest type of chemical equilibrium and corresponds to the equilibrium between two isomers such as n-butane and isobutane.† If dn_A

---

† n-butane $CH_3-CH_2-CH_2-CH_3$, isobutane $CH_3-CH-CH_3$.
$\qquad\qquad\qquad\qquad\qquad\qquad\qquad\qquad\qquad\quad |$
$\qquad\qquad\qquad\qquad\qquad\qquad\qquad\qquad\quad CH_3$

moles of A are converted into dn_B moles of B at constant T and P we have dG $= (+\mu_A dn_A + \mu_B dn_B)$, where dn_A is negative and dn_B positive. We can define an *extent of reaction* ξ which is 0 when the reaction position is entirely to the left of the equation (i.e. only reactants present) and is 1 when one mole of reactant has gone over entirely to products. In our simple example we may write

$$d\xi = dn_B = -dn_A,$$

and

$$dG = (\mu_B - \mu_A)d\xi \quad \text{at constant } T \text{ and } P.$$

The reaction will proceed until G reaches a minimum value and $\left(\dfrac{\partial G}{\partial \xi}\right)_{T,P}$ $= 0$, as illustrated in Fig. 4.8. As $\left(\dfrac{\partial G}{\partial \xi}\right)_{T,P} = \mu_B - \mu_A$, this is the position where $\mu_A = \mu_B$.

If the components follow the perfect-gas laws,

$$\mu_i = \mu_i^0 + RT \ln(P_i/\text{atm}) \qquad \text{(Section 4.10)}.$$

Therefore

$$\left(\frac{\partial G}{\partial \xi}\right)_{T,P} = \mu_B - \mu_A = \mu_B^0 - \mu_A^0 + RT \ln \frac{P_B}{P_A}.$$

$\mu_B^0 - \mu_A^0$ is ΔG^0, the free-energy change if one mole of reaction takes place with both reactant and product remaining in their standard state (at 1 atm pressure).

$$\left(\frac{\partial G}{\partial \xi}\right)_{T,P} = \Delta G^0 + RT \ln \frac{P_B}{P_A}.$$

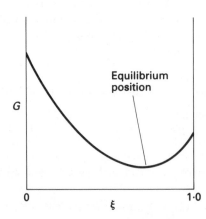

Fig. 4.8. The Gibbs free energy as a function of the extent of a chemical reaction (ξ).

$\left(\dfrac{\partial G}{\partial \xi}\right)_{T,P}$ is the change in free energy with the degree of advancement of the reaction under the conditions specified by P_A and P_B. It is equal to the free-energy change for one mole of reaction with A at partial pressure P_A going to B at partial pressure P_B, the partial pressures remaining constant. In modern treatments $-\left(\dfrac{\partial G}{\partial \xi}\right)_{T,P}$ is called the *affinity* of the reaction whereas in older or elementary texts $\left(\dfrac{\partial G}{\partial \xi}\right)_{T,P}$ is called the *reaction free energy* and written simply as ΔG. We will write it as $\Delta G'$, the prime reminding us that it is in reality a differential quantity and only corresponds to the free-energy change for a mole of reaction under precisely defined conditions.

At equilibrium

$$\Delta G' = \left(\frac{\partial G}{\partial \xi}\right)_{T,P} = 0,$$

and as

$$\Delta G' = \Delta G^0 + RT \ln \frac{P_B}{P_A},$$

$$\Delta G^0 = -RT \ln \left(\frac{P_B}{P_A}\right)_{eq}.$$

We call the value of $\left(\dfrac{P_B}{P_A}\right)$ at equilibrium the equilibrium constant of the reaction, K_P.

$$\Delta G^0 = -RT \ln K_P.$$

This important equation tells us how the position of chemical equilibrium can be defined in terms of the free energies of the reactants and products at 1 atm pressure. Such standard free energies can be determined experimentally and are tabulated for use in this way. We shall consider specific examples later. The equation is also valuable in a qualitative sense. If ΔG^0 is negative we know the equilibrium position will correspond to the presence of more product than reactants ($\ln K_P > 0$). If ΔG^0 is positive the reaction will not proceed to such an extent and reactants will predominate in the equilibrium mixture. With this result we have accomplished a major purpose of our study.

Had our reaction been more complicated we would have obtained essentially the same results. For example for the reaction

$$a\mathrm{A} + b\mathrm{B} \rightleftharpoons l\mathrm{L} + m\mathrm{M}$$

Table 4.2 Standard free energy changes and the corresponding values of the equilibrium constant at 298 K

$\Delta G^{\circ}/\text{kJ mol}^{-1}$	K	Composition of equilibrium mixture
− 50	6×10^8	negligible reactants
− 10	57	products dominate
− 5	7.5	
0	1.0	
+ 5	0.13	reactants dominate
+ 10	0.02	
+ 50	1.7×10^{-9}	negligible products

we would have

$$dG = \mu_L dn_L + \mu_M dn_M + \mu_A dn_A + \mu_B dn_B = \Sigma \mu_i dn_i.$$

The extent of reaction ξ would have to be defined in a more complicated manner so that

$$d\xi = \frac{dn_L}{l} = \frac{dn_M}{m} = -\frac{dn_A}{a} = -\frac{dn_B}{b} = \frac{dn_i}{\nu_i}$$

where ν_i represents the stoichiometric coefficients $-a$, $-b$, m and l. The ν_i for the reactants are defined as negative and those for the products as positive. Thus

$$dG = (l\mu_L + m\mu_M - a\mu_A - b\mu_B)d\xi$$

and

$$\Delta G' = \left(\frac{\partial G}{\partial \xi}\right)_{T,P} = (l\mu_L + m\mu_M - a\mu_A - b\mu_B) = \sum_i \nu_i \mu_i.$$

As $\mu_A = \mu_A^0 + RT \ln (P_A/\text{atm})$ etc., we obtain

$$\Delta G' = \Delta G^0 + RT \ln \left[\frac{(P_L/\text{atm})^l (P_M/\text{atm})^m}{(P_A/\text{atm})^a (P_B/\text{atm})^b}\right]$$

which we can write in more concise notation

$$\Delta G' = \Delta G^0 + RT \ln \prod_i (P_i/\text{atm})^{\nu_i}$$

At equilibrium $\Delta G^0 = -RT \ln K_P$, where

$$K_P = \left[\frac{(P_L/\text{atm})^l (P_M/\text{atm})^m}{(P_A/\text{atm})^a (P_B/\text{atm})^b}\right]_{\text{eq}} = \left[\prod_i (P_i/\text{atm})^{\nu_i}\right]_{\text{eq}}$$

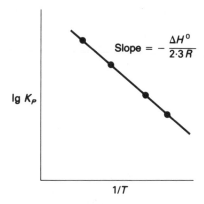

Fig. 4.9. The logarithm of an equilibrium constant as a function of reciprocal temperature (schematic).

and

$$\Delta G^0 = \sum_i v_i \mu_i^0,$$

the standard free-energy change for a mole of reaction.

These equations, though they look a little more complicated, are fundamentally the same as those we obtained in our simple illustrative example of butane isomerization. K_P is strictly dimensionless even if $(a + b) \neq (l + m)$, as all the pressures comprising it are themselves dimensionless ratios, i.e. $P(\text{atm})/(1 \text{ atm})$ (Section 4.3).

4.12 Temperature-dependence of equilibrium constants

We can use the thermodynamic relations we have obtained so far to find out how the position of equilibrium will change if we alter the temperature.

Remembering that using

$$\left(\frac{\partial G}{\partial T}\right)_P = -S \qquad \text{(Section 4.4).}$$

and

$$G = H - TS \qquad \text{(Section 4.2),}$$

we obtained the Gibbs–Helmholtz equation (Section 4.4)

$$\left[\frac{\partial\left(\frac{\Delta G}{T}\right)}{\partial T}\right]_P = -\frac{\Delta H}{T^2},$$

we can differentiate the equation $\Delta G^0 = -RT \ln K_P$ (Section 4.11) obtaining

$$\left(\frac{\partial \ln K_P}{\partial T}\right) = -\left[\frac{\partial\left(\frac{\Delta G^0}{T}\right)}{\partial T}\right]_P = \frac{\Delta H^0}{RT^2}.$$

This important equation

$$\left(\frac{\partial \ln K_P}{\partial T}\right)_P = \frac{\Delta H^0}{RT^2}$$

is called the Van't Hoff Isochore. As ΔG^0 and ΔH^0 are not functions of pressure (as they are by definition the values at 1 atm) we can write

$$\frac{\mathrm{d} \ln K_P}{\mathrm{d}T} = \frac{\Delta H^0}{RT^2}.$$

If we assume ΔH^0 is independent of temperature (which is often a fair approximation) then integration gives

$$\ln\frac{K_2}{K_1} = -\frac{\Delta H^0}{R}\left(\frac{1}{T_2} - \frac{1}{T_1}\right).$$

Plotting $\lg K_P$ against $1/T$, as illustrated in Fig. 4.9, the slope is $-\Delta H/2.3R$. For an exothermic reaction ($\Delta H < 0$) K_p must decrease as the temperature increases. Thus with the reaction

$$N_2 + 3H_2 \rightleftharpoons 2NH_3; \qquad \Delta H^0 = -92.4\,\mathrm{kJ\,mol^{-1}}$$

we predict and indeed obtain less ammonia in the equilibrium mixture at higher temperatures. For an endothermic reaction ($\Delta H > 0$) K_p increases with increasing temperature; so for the equilibrium

$$N_2O_4 \rightleftharpoons 2NO_2; \qquad \Delta H^0 = 58.0\,\mathrm{kJ\,mol^{-1}}$$

we predict and observe more dissociation at higher temperatures.

Example

The equilibrium constant K_P for the dissociation of bromine into atoms

$$Br_2 \rightleftharpoons Br\cdot + Br\cdot$$

is 6×10^{-12} at 600 K and 1×10^{-7} at 800 K. Calculate the standard free energy change for the reaction at these temperatures and the standard enthalpy change assuming this is constant in the temperature range 600–800 K.

At 600 K

$$\Delta G^0 = -RT \ln K_P \quad \text{(Section 4.11)}$$

$$\Delta G^0 = -8.3 \times 600 \times \ln(6 \times 10^{-12}) \,\text{J}$$

$$\Delta G^0 = 129 \,\text{kJ}\,\text{mol}^{-1} \text{ at } 600 \,\text{K}.$$

At 800 K

$$\Delta G^0 = -8.3 \times 800 \times 2.3 \times (-7) \,\text{J}$$

$$\Delta G^0 = 107 \,\text{kJ}\,\text{mol}^{-1} \text{ at } 800 \,\text{K}$$

$$\frac{d \ln K_P}{dT} = \frac{\Delta H^0}{RT^2} \quad \text{(Section 4.12)}.$$

Integrating

$$\ln \frac{K_P(T_2)}{K_P(T_1)} = \frac{\Delta H^0}{R} \left(\frac{1}{T_1} - \frac{1}{T_2} \right)$$

$$\frac{\Delta H^0}{8.3} \left(\frac{1}{600} - \frac{1}{800} \right) = \ln \left(\frac{10^{-7}}{6 \times 10^{-12}} \right)$$

$$\Delta H^0 = 193 \,\text{kJ}\,\text{mol}^{-1}.$$

We can also calculate the standard entropy change associated with the reaction as $\Delta G^0 = \Delta H^0 - T \Delta S^0$.

At 600 K

$$\Delta S^0 = \frac{(193 - 129)}{600} \times 10^3 \,\text{J}\,\text{K}^{-1}\,\text{mol}^{-1} = 107 \,\text{J}\,\text{K}^{-1}\,\text{mol}^{-1}.$$

As we would expect, the energy of the molecules is large and negative with respect to the constituent atoms, thus the enthalpy change on dissociation is positive and unfavourable to the dissociation process. However, the additional freedom acquired by the dissociated atoms leads to a positive entropy change. This favours dissociation, which becomes more significant as the temperature is increased.

4.13 Effect of pressure on equilibrium constants

As K_P may be defined in terms of the free energies of the participant substances in the *standard states*, i.e. at 1 atm pressure, it must be independent of pressure.

Consider the reaction

$$A_2(g) \rightleftharpoons 2A(g)$$
$$1 - \alpha \quad\quad 2\alpha$$

$$K_P = \frac{(P_A/\text{atm})^2}{(P_{A_2}/\text{atm})} \quad \text{(Section 4.11)}$$

If we express the partial pressures in terms of the degree of dissociation α,

$$P_A = \frac{2\alpha}{1 + \alpha} P \qquad \text{(Section 1.5)}$$

and

$$P_{A_2} = \frac{1 - \alpha}{1 + \alpha} P$$

where P is the total pressure. Therefore

$$K_P = \frac{4\alpha^2 (P/\text{atm})}{(1 - \alpha)(1 + \alpha)} = \frac{4\alpha^2 (P/\text{atm})}{(1 - \alpha^2)}.$$

Thus as K_P is independent of the total pressure, the degree of dissociation α must in fact decrease as P increases.

Example

At 298 K the equilibrium constant for the dissociation

$$N_2O_4 \rightleftharpoons 2NO_2$$

is 0.14. Calculate the fraction of N_2O_4 molecules dissociated at 1 atm and 10 atm pressure.

As

$$K_P = \frac{4\alpha^2 (P/\text{atm})}{1 - \alpha^2}$$

we obtain

$$\alpha = \left(\frac{K_P}{4P + K_P}\right)^{\frac{1}{2}}.$$

Substituting $K_P = 0.14$ and $P = 1$ we obtain

$$\alpha = \left(\frac{0.14}{4.14}\right)^{\frac{1}{2}} = 0.18.$$

At 10 atm,

$$\alpha = \left(\frac{0.14}{40.14}\right)^{\frac{1}{2}} = 0.06.$$

If K_P is small compared with P/atm, α is approximately proportional to $(1/P)^{\frac{1}{2}}$.

If we can express the equilibrium constant in terms of mole fractions defined by

$$x_i = \frac{n_i}{n}$$

where n is the total number of moles in the system, then for the equilibrium

$$aA + bB \rightleftharpoons lL + mM,$$

$$K_x = \frac{x_L^l\, x_M^m}{x_A^a\, x_B^b}.$$

As the partial pressure is proportional to the mole fraction for perfect-gas mixtures, $(P_i/\text{atm}) = x_i(P/\text{atm})$, and so

$$K_x = \frac{(P_L/\text{atm})^l\,(P_M/\text{atm})^m}{(P_A/\text{atm})^a\,(P_B/\text{atm})^b} \cdot (P/\text{atm})^{(a+b-l-m)}$$

or

$$K_x = K_P(P/\text{atm})^{-\Delta n}$$

where $\Delta n = l + m - a - b$ is the change in the number of moles of gaseous substances, as the reaction goes from left to right. Thus unless $\Delta n = 0$ the mole fractions of the components of the equilibrium mixture will depend on the total pressure even though K_P does not. If the reaction is such as to increase the number of moles of gas, an increase in pressure will reduce the mole fraction of products in the final equilibrium mixture. We can express the dependence of K_x on pressure more generally.

$$\left(\frac{\partial \ln K_x}{\partial P}\right)_T = \left(\frac{\partial \ln K_P}{\partial P}\right)_T - \Delta n \left(\frac{\partial \ln P}{\partial P}\right)_T.$$

For perfect gases $P\Delta V = \Delta nRT$. Therefore as $\left(\dfrac{\partial \ln K_P}{\partial P}\right)_T = 0$, we obtain

$$\left(\frac{\partial \ln K_x}{\partial P}\right)_T = -\frac{\Delta n}{P} = -\frac{\Delta V}{RT}.$$

This equation applies not only to equilibria involving gases but also to equilibria in solution, and indeed to any equilibrium when the equilibrium constant is expressed in terms of mole fractions rather than partial pressures. In these circumstances ΔV is the volume change accompanying one mole of reaction.†

4.14 Basic results of chemical thermodynamics

A famous physical chemist, G. N. Lewis, once stated: 'Thermodynamics exhibits no curiosity'. By this he meant that the conclusions of thermo-

---

† Strictly, one mole of reaction with all substances in their standard states. These new standard states, to be discussed later, are defined at unit concentrations rather than at unit partial pressures and are not the standard states we have used up to the present.

dynamics are quite general. Thus if we apply thermodynamics to an equilibrium we will get the correct answer even if, for example, we have an entirely wrong idea of the nature of the molecules making up our system. For this reason there can be relatively few important thermodynamics equations and these must be of extremely wide applicability. You may have noticed that when studying both the vapour pressure of a liquid and a chemical reaction we obtained equations which, though known by different names, had exactly the same form. We shall now summarize these equations.

(i) For any system at equilibrium $\Delta G^\ominus = -RT \ln K$ (Section 4.11), where K is some quantity which characterizes the equilibrium position in terms of the amounts of materials present in the equilibrium mixture. Thus K could be an equilibrium constant, a vapour pressure, or a solubility. $\Delta G^\ominus$ is a standard free-energy change for one mole of reaction for the equation which describes the equilibrium. (In this general formulation the standard states are not necessarily those based on the 1 atm standard.)

(ii) The second type of equation is $\left(\dfrac{\partial \ln K}{\partial T}\right)_P = \dfrac{\Delta H^\ominus}{RT^2}$ (Section 4.12), where again K is a quantity which characterizes the position of equilibrium and $\Delta H^\ominus$ is the standard enthalpy change for one mole of reaction.

(iii) The effect of pressure on the position of equilibrium may be expressed as $\left(\dfrac{\partial \ln K_x}{\partial P}\right)_T = -\dfrac{\Delta V^\ominus}{RT}$ (Section 4.13), where $\Delta V^\ominus$ is the volume change accompanying a mole of reaction under standard conditions.

(iv) The final equation we recall, Section 4.11, can be written in the generalized form

$$\Delta G' = \left(\frac{\partial G}{\partial \xi}\right)_{T,P} = \Delta G^\ominus + RT \ln \prod_i \left(\frac{c_i}{c^\ominus}\right)^{v_i}$$

where $\left(\dfrac{c_i}{c^\ominus}\right)$ is some measure of the quantity of the ith constituent of the reaction mixture relative to that in the standard state, $c^\ominus$.

These equations depend on the fact that

$$\mu_i = \mu_i^\ominus + RT \ln \left(\frac{c_i}{c^\ominus}\right)$$

where $\left(\dfrac{c_i}{c^\ominus}\right)$ represents some suitable measure of concentration of i, for example, (P_i/atm) or x_i. $\mu_i^\ominus$ is the chemical potential of the standard state, that is the state for which $c_i = c^\ominus$. If we select the one atmosphere standard state we used above when considering reactions in the gaseous phase, $\mu^\ominus = \mu^0$.

4.15 Le Chatelier's Principle

The direction of change in the position of equilibrium due to change in the external variables such as T and P may generally be found by the application of Le Chatelier's Principle which states: *Perturbation of a system at equilibrium will cause the equilibrium position to change in such a way as to tend to remove the perturbation.*

For instance, if heat is evolved in a reaction ($\Delta H < 0$), lowering the temperature will promote more reaction because moving the equilibrium toward the product side tends to raise the temperature of the system. If a reaction proceeds with a positive volume change, then application of pressure shifts the equilibrium in the direction of the reactants. These conclusions are given quantitative expression in the equations

$$\left(\frac{\partial \ln K}{\partial T}\right)_P = \frac{\Delta H^0}{RT^2} \quad \text{(Section 4.12);}$$

$$\left(\frac{\partial \ln K_x}{\partial P}\right)_T = -\frac{\Delta V}{RT} \quad \text{(Section 4.13).}$$

Le Chatelier's Principle provides a good guide to the effects of pressure and temperature changes. To make it universally true however it would have to be stated in a more rigorous form, so it is wise to regard it as a useful guide or *aide memoire* rather than as a fundamental thermodynamic principle.

Problems

4.1. Naphthalene melts at 353 K at 1 atm pressure with an enthalpy change on fusion of $19\,kJ\,mol^{-1}$. The volume increase on fusion is $19 \times 10^{-6}\,m^3$. What change in the melting temperature will be observed if the pressure is raised by 100 atm? ($1\,atm = 10^5\,N\,m^{-2}$.)

4.2. The vapour pressure of a liquid is

T/K	326.1	352.9	415.0	451.7
Vapour pressure/mm Hg	1.0	5.0	100	400

Calculate the enthalpy change on vaporization and the entropy of vaporization. Calculate the standard Gibbs free-energy change accompanying vaporization at 373 K.

4.3. Formic acid is partially associated into dimers in the vapour phase. The mole fraction present as monomer is 0.228 at 283 K and 10 mm Hg pressure and 0.715 at 333 K and 16 mm Hg. Calculate the enthalpy change on dimerization. ($760\,mmHg = 1\,atm = 10^5\,N\,m^{-2}$.)

4.4. In a mixture at 1 atm the partial pressure of CO in equilibrium with CO_2 and C was as follows:

T/K	1083	1173	1253
Partial pressure of CO/atm	0.931	0.978	0.991

Calculate the enthalpy change accompanying the reaction $CO_2 + C \rightarrow 2CO$.

4.5. The equilibrium constant for the reaction $H_2 + I_2 \rightleftharpoons 2HI$ is 45.6 at 764 K and 60.8 at 667 K. Estimate the enthalpy change which accompanies one mole of the forward reaction.

4.6. The enthalpy of vaporization of water at 373 K and 1 atm is $40.7\,kJ\,mol^{-1}$. Estimate the vapour pressure of water at 368 K.

4.7. In the *Handbook of Chemistry and Physics* the following data are given for the vapour pressure of mercury

Temp (°C)	0	4	8	12
Pressure (mmHg)	0.000185	0.000276	0.000406	0.000588

Calculate the entropy and enthalpy changes accompanying vaporization at 1 atm and the Gibbs free energy change accompanying vaporization at 273 K and at the boiling point of mercury.

4.8. For the gas phase reaction

$$\text{cyclopentene} + I_2 = \text{cyclopentadiene} + 2HI$$

$$\log K_p = 7.55 - 4817\, T^{-1} \text{ in the temperature range } 450\text{–}690\,K.$$

Calculate the standard enthalpy, entropy, and Gibbs free energy changes accompanying the reaction at 573 K.

5
Determination of thermodynamic quantities

5.1 Hess's Law

The equations we have obtained will be of value to us for any process only if we can readily evaluate ΔG^0, ΔH^0, and ΔS^0. The fact that we are dealing with state functions is of considerable help, as the changes in a state function summed over a complete cycle must be zero, and the change in any state function between two states is constant and independent of the path taken between the states (Section 2.6). This principle was first stated by Hess (1840) with specific reference to enthalpy changes. Its value lies in the fact that the enthalpy changes accompanying some reactions are easy to measure whereas others are difficult.

As an example let us consider the enthalpy change when methane is formed from its elements, both reactants and product being in their standard states. This is called the standard enthalpy of formation and written ΔH_f^0.

$$C(s) + 2H_2(g) \rightarrow CH_4(g).$$

The standard state of each element is defined as the most stable form at 1 atm and the temperature specified (most frequently enthalpies of formation are measured and quoted at 298 K). The direct reaction cannot conveniently be carried out, but it is relatively easy to measure the enthalpy of combustion of methane in an apparatus called a flame calorimeter. As $\Delta H = (q)_P$ (Section 2.7), when methane is burnt with oxygen the heat produced gives the enthalpy of combustion directly.

$$CH_4(g) + 2O_2(g) \rightarrow CO_2(g) + 2H_2O(l); \qquad \Delta H = -890.4 \text{ kJ mol}^{-1}$$
$$C(s) + O_2(g) \rightarrow CO_2(g); \qquad \Delta H = -393.5 \text{ kJ mol}^{-1}$$
$$2H_2(g) + O_2(g) \rightarrow 2H_2O(l); \qquad \Delta H = -571.6 \text{ kJ mol}^{-1}.$$

Combining the enthalpy of combustion of methane with that of carbon and hydrogen enables us to apply Hess's Law to the problem. We can see from Fig. 5.1 that

$$\Delta H_f^0 = \Delta H_1 - \Delta H_2.$$

ΔH_1 is the enthalpy of combustion of one mole of carbon and two moles of

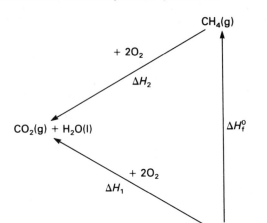

Fig. 5.1. The heat of formation of methane from its elements and the two combustion processes which taken together are equivalent.

hydrogen. Therefore

$$\Delta H_f^0(CH_4) = [-393.5 - 571.6 + 890.4] \, kJ \, mol^{-1}$$
$$= 74.7 \, kJ \, mol^{-1}$$

It is not always necessary to draw out the alternative paths. If we write out a series of chemical reactions together with the enthalpy changes that accompany them, then addition gives the enthalpy change corresponding to the overall reaction. The overall reaction and the constituent reactions taken together represent two different ways of carrying out the same process.

$$\Delta H/kJ \, mol^{-1}$$

$$C(s) + O_2(g) \rightarrow CO_2(g) \qquad -393.5$$
$$2H_2(g) + O_2(g) \rightarrow 2H_2O(l) \qquad -571.6$$
$$CO_2(g) + 2H_2O(l) \rightarrow CH_4(g) + 2O_2(g) \qquad +890.4$$

Adding,

$$2H_2(g) + C(s) \rightarrow CH_4(g) \qquad -74.7.$$

If the reverse chemical reaction is considered, the enthalpy change reverses its sign.

5.2 Standard enthalpies of formation

Standard enthalpies of formation of chemical compounds† are of great practical value to chemists. They give an easy method of determining the

---

† Remember: the enthalpy change when a compound is formed from its elements in their standard states, i.e. normal states of matter at 1 atm pressure, ΔH_f^0, are usually tabulated for 298 K.

enthalpy change accompanying any reaction, as

$$\Delta H^0 = \Sigma \Delta H_f^0(\text{products}) - \Sigma \Delta H_f^0(\text{reactants}).$$

The validity of the equation is based on the fact that when chemical reactions occur the total elemental composition is unchanged—the elements are merely redistributed with different partners. We can confirm this by considering the reactions

$$CO(g) + \tfrac{1}{2}O_2(g) \rightarrow CO_2(g)$$

and

$$C(s) + \tfrac{1}{2}O_2(g) \rightarrow CO(g); \qquad \Delta H_f^0[CO]$$
$$C(s) + O_2(g) \rightarrow CO_2(g); \qquad \Delta H_f^0[CO_2].$$

If we write

$$C(s) + O_2(g) \rightarrow CO_2(g); \qquad \Delta H_f^0[CO_2]$$
$$CO(g) \rightarrow C + \tfrac{1}{2}O_2(g); \qquad -\Delta H_f^0[CO].$$

Adding,

$$CO(g) + \tfrac{1}{2}O_2(g) \rightarrow CO_2(g),$$

and

$$\Delta H^0 = \Delta H_f^0[CO_2] - \Delta H_f^0[CO]$$

in keeping with the general equation.

The enthalpies of formation of elements in their standard states are of course zero (the formation of an element from its elements!). If the element is not in its standard state then we would have to make allowance for this. Thus if we were considering a reaction involving bromine in the gaseous state we would have to allow for its enthalpy of vaporization.

5.3 Average bond energies

The enthalpy change for the dissociation of a diatomic molecule such as H_2 into its atoms in the gas phase can be termed the *bond dissociation energy* (or more strictly the bond dissociation enthalpy). If we consider the reaction

$$CH_4(g) \rightarrow C(g) + 4H(g)$$

we could identify the enthalpy change accompanying this reaction with the energy of dissociation of four C–H bonds. This enthalpy change can be directly calculated from the following thermodynamic data.

$$CH_4(g) \rightarrow C(s) + 2H_2(g); \qquad \Delta H = -\Delta H_f^0[CH_4] = +75 \text{ kJ mol}^{-1}$$
$$C(s) \rightarrow C(g); \qquad \Delta H = +\Delta H_f^0[C(g)] = +717 \text{ kJ mol}^{-1}$$
$$2H_2(g) \rightarrow 4H(g); \qquad \Delta H = 4\Delta H_f^0[H(g)] = +872 \text{ kJ mol}^{-1}$$
$$CH_4(g) \rightarrow C(g) + 4H(g); \qquad \Delta H = 1664 \text{ kJ mol}^{-1}.$$

Table 5.1 Average values of bond energies/$(kJ\ mol^{-1})$

H—H	436	C—C	348	C=C	615
H—C	414	C—N	292	C≡C	812
H—O	463	C—O	351	C=O	728
H—Cl	431	C—F	255	C≡N	879
		C—Cl	343		
		C—Br	289		

Thus the average bond dissociation energy of the carbon–hydrogen bond may be estimated at $416\ kJ\ mol^{-1}$. Similar thermodynamic exercises enable us to construct a set of bond energies, as given in Table 5.1, which we can use to estimate the enthalpy changes accompanying reactions. The method is only approximate, as indeed is the concept of a specific energy associated with a bond between two elements. In practice the energy associated with a chemical bond may vary according to its environment in molecules.

For example the enthalpy of formation of ethane may be calculated from bond dissociation energies ε as follows.

$$2C(s) + 3H_2(g) \rightarrow C_2H_6(g)$$

$$-\Delta H_f^0[C_2H_6] = 1\varepsilon_{C-C} + 6\varepsilon_{C-H} - 3\varepsilon_{H-H} - 2\Delta H_{vap}[C]$$

$$= (348 + 2484 - 1308 - 1434)\ kJ\ mol^{-1}.$$

Therefore

$$\Delta H_f^0 = -90\ kJ\ mol^{-1}.$$

The experimental value for the enthalpy of formation of ethane is $-84.5\ kJ\ mol^{-1}$ and the value calculated from the difference of very large energies is only in approximate agreement.

5.4 Temperature-dependence of enthalpy changes

As

$$\left(\frac{\partial H}{\partial T}\right)_P = C_P$$

then

$$\left(\frac{\partial(\Delta H)}{\partial T}\right)_P = \Delta C_P.$$

This is called *Kirchhoff's equation*; ΔH is the enthalpy change accompanying a reaction and ΔC_P is the sum of the heat capacities of the products less the sum of the heat capacities of the reactants. Integrating from temperature T_1

to T_2,

$$\Delta H_2 - \Delta H_1 = \int_{T_1}^{T_2} \Delta C_P \, dT.$$

If the difference in heat capacities is independent of temperature,

$$\Delta H_2 - \Delta H_1 = \Delta C_P (T_2 - T_1).$$

More generally the heat capacities of the components can be expressed in the form

$$C_P = a + bT + cT^2 \ldots,$$

and the integral evaluated in the normal way.

Example

The enthalpy change ΔH when water freezes at 273 K is $-6.00 \text{ kJ mol}^{-1}$. C_P for water is $75.3 \text{ J K}^{-1} \text{mol}^{-1}$ and for ice $37.6 \text{ J K}^{-1} \text{mol}^{-1}$. Calculate the enthalpy change when water freezes at 253 K.

$$H_2O(l) \rightarrow H_2O(s)$$
$$\Delta H_2 = \Delta H_1 + \Delta C_p (T_2 - T_1)$$
$$\Delta H_2 = -6000 + (37.6 - 75.3)(253 - 273) \text{ J mol}^{-1}$$
$$\Delta H_2 = -6000 + 754 \text{ J mol}^{-1}$$

or

$$\Delta H_2 = -5.2 \text{ kJ mol}^{-1}.$$

5.5 Standard free energies of formation

Standard Gibbs free energies of formation of compounds from their elements may be defined in the same way as enthalpies.

$$\Delta G^0 = \Sigma \Delta G_f^0 (\text{products}) - \Sigma \Delta G_f^0 (\text{reactants}).$$

As we have seen (Section 4.11), ΔG^0 defines the equilibrium position for a chemical reaction as $\Delta G^0 = -RT \ln K_p$. For a reaction such as

$$A + B \rightleftharpoons C + D$$

if ΔG^0 is negative K_p will be greater than unity and the products will predominate. However, if ΔG^0 is positive K_p will be less than unity and the reactants will predominate in the equilibrium mixture. Though such information is a useful guide to the feasibility of carrying out a chemical reaction it may sometimes be misleading. Thus at 298 K for the reaction

$$H_2(g) + \tfrac{1}{2}O_2(g) \rightarrow H_2O(l); \qquad \Delta G^0 = -237.2 \text{ kJ mol}^{-1}.$$

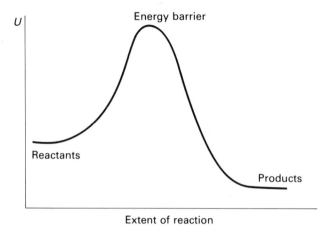

Fig. 5.2.   Energy barrier between the reactants and the products of a chemical reaction.

The high negative value of ΔG^0 indicates that very little hydrogen and oxygen would be present in the equilibrium state. Yet a mixture of H_2 and O_2 can be kept for a long time in the absence of a suitable catalyst (or a spark) without the formation of water. Thus, though ΔG^0 tells us which state of the system is the most stable from a thermodynamic standpoint, this state is not automatically attained by the system. Frequently, in passing from one thermodynamic state to another, as in a chemical reaction, there may be a high energy barrier between the two states as illustrated in Fig. 5.2. Thus the *rate* of passage to the state of lowest free energy may be so slow that the equilibrium state will not be attained in any reasonable time.

5.6   Determination of free energy changes

ΔG^0 may of course be determined from observed equilibrium constants. However, we often wish to be able to do the opposite and predict equilibrium constants from standard free energies. To do this we make use of the equation (Section 4.2)

$$\Delta G^0 = \Delta H^0 - T\Delta S^0.$$

For a reaction ΔH^0 can be measured either directly by the use of calorimetry or indirectly by making use of Hess's Law. For any reaction we also have

$$\Delta S^0 = \Sigma \Delta S_f^0(\text{products}) - \Sigma \Delta S_f^0(\text{reactants}),$$

so that a knowledge of standard entropies of formation would enable us to determine ΔS^0. However we are able to determine ΔS^0 by a more direct method.

5.7 Determination of entropies of substances

As we have already established the entropy of a substance can be determined from the equations

$$dS = \frac{dq_{rev}}{T} \quad \text{(Section 3.3)} \quad \text{and} \quad C_P = \left(\frac{dq}{dT}\right)_P \quad \text{(Section 2.8)},$$

which give at constant pressure

$$dS = \frac{C_P}{T} dT,$$

$$S_T - S_0 = \int_0^T \frac{C_P}{T} dT \quad \text{(Section 3.9)}.$$

However, unlike enthalpy and free energy, the *absolute* entropy of a substance can be determined by invoking the Third Law of Thermodynamics. This may be expressed: *The entropy of all perfect crystals is zero at the absolute zero of temperature.* Most pure substances form essentially perfect crystals at low temperatures and in such cases we can assume $S_0 = 0$. Therefore

$$S_T = \int_0^T \frac{C_P}{T} dT = \int_0^T C_P \, d(\ln T).$$

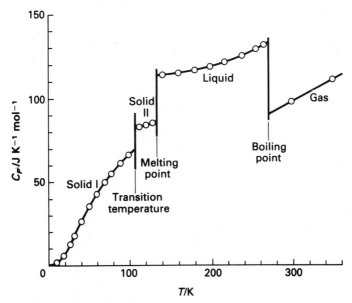

Fig. 5.3. The heat capacity at constant pressure (C_P) of n-butane as a function of temperature (T).

From the measurement of C_P as a function of T (Fig. 5.3), S_T may be evaluated by integration. If the substance is other than solid at temperature T some heat will be absorbed isothermally at the melting point (and, if it is a gas, at the boiling point as well). The contribution of these isothermal processes must be added to the integral.

$$\Delta S_{\mathrm{fus}} = \frac{\Delta H_{\mathrm{fus}}}{T_{\mathrm{fus}}},$$

and

$$\Delta S_{\mathrm{vap}} = \frac{\Delta H_{\mathrm{vap}}}{T_{\mathrm{vap}}}.$$

Thus the absolute entropies of elements and compounds can be established. These can be used to determine the entropy changes accompanying chemical reactions.

$$\Delta S^0 = \Sigma S^0(\text{products}) - \Sigma S^0(\text{reactants}).$$

The basis for the Third Law can be explained in molecular terms. At absolute zero all matter will be in the configuration that has the lowest possible energy. This occurs when all the molecules are in the state of lowest energy. The number of 'arrangements' of the system that satisfies this condition is only one, thus

$$S_0 = k \ln W = k \ln 1 = 0 \qquad \text{(Section 3.10)}.$$

Sometimes molecules get 'frozen in' to other states so that the perfect crystal state, the true equilibrium state, is not attained at low temperatures. Then S_0 as measured is not zero. Such cases are often, but misleadingly, called 'exceptions to the Third Law'. Thus a number of substances form glassy solids which remain apparently stable at low temperatures despite the fact that the crystalline state is the one of lowest free energy. This is because it would take an extremely long time, at low temperatures, for the molecules of the glass to rearrange themselves to the pattern required for crystallization.

5.8 Example of the determination of thermodynamic quantities

To see how the methods of determining thermodynamic quantities are applied let us return to the simple equilibrium we have used in earlier examples,

$$\text{n-}C_4H_{10} \rightleftharpoons \text{i-}C_4H_{10}.$$

To determine K_P from thermodynamic data for the two isomers we need ΔG^0. In order to calculate ΔG^0 we must obtain ΔH^0 and ΔS^0. These

Table 5.2 The determination of K_P from thermodynamic measurements

To determine	We need	Since they are related by	
K_P	ΔG^0	$\Delta G^0 = -RT \ln K_P$	(4.11)
ΔG^0	$\Delta H^0, \Delta S^0$	$\Delta G^0 = \Delta H^0 - T\Delta S^0$	(4.2)
ΔH^0	ΔH_f^0	$\Delta H^0 = \Sigma \Delta H_f^0$ (products) $- \Sigma \Delta H_f^0$ (reactants)	(5.2)
ΔH_f^0	ΔH^0 (combustion)	Hess's Law	(5.1)
ΔH^0 (combustion)	We must measure		
ΔS^0	S^0	$\Delta S^0 = \Sigma S^0$ (products) $- \Sigma S^0$ (reactants)	(5.6)
S^0	C_P	$S^0 = S_0^0 + \displaystyle\int_0^T \frac{C_P}{T}\,dT$	(5.7)
C_P	We must measure		

quantities can be calculated from the ΔH_f^0 and S^0 of the isomers. ΔH_f^0 is best determined from enthalpy changes on combustion and S^0 requires a knowledge of the heat capacities from very low temperatures up to the temperature of interest. So, starting from the practical end, let us first find out how to determine the enthalpy change on combustion and C_P for both isomers.

Bomb calorimeter

The enthalpy changes on combustion of solids and liquids are determined by using a bomb, which in thermochemistry is merely a strong metal vessel immersed in a water jacket (Fig. 5.4). A small quantity of the experimental substance is placed in the bomb which is then filled with excess oxygen under pressure (20–30 atm). The material is ignited and the rise in temperature of the water jacket noted. The apparatus is calibrated using a substance whose enthalpy change on combustion is known, or by producing an equivalent temperature rise in the jacket by electrical heating. In this way the change in temperature of the jacket can be directly related to the energy change on combustion for the substance under investigation. Needless to say, to obtain accurate values many experimental refinements and theoretical corrections to the results must be applied. As the measurements in the bomb are performed at constant volume it is in fact ΔU and not ΔH which is measured ($\Delta U = q_V$). The correction $\Delta H = \Delta U + \Delta(PV) = \Delta U + \Delta nRT$, where Δn is the change in the number of moles of gases accompanying the combustion, is not difficult.

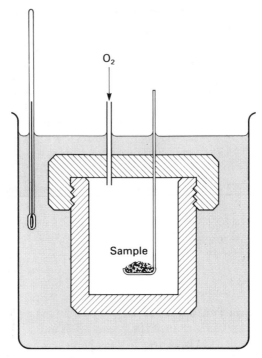

Fig. 5.4. Bomb calorimeter.

Flame calorimeter

The enthalpy changes on combustion of gases, such as the butane isomers, are usually determined in a flame calorimeter (Fig. 5.5). The gas is burnt in a flame with excess oxygen and the heat produced is measured by observing the rate at which the combustion products heat a known quantity of water. The apparatus must be calibrated using electrical heating. This apparatus, working at constant pressure, gives ΔH directly from the heat evolved on combustion. The results of such an experiment give -2877.13 and $-2868.76 \text{ kJ mol}^{-1}$ for the standard enthalpy changes on combustion of n-butane and of isobutane respectively. ΔH_f^0 can be calculated by the following scheme;

$$C_4H_{10}(g) + 6\tfrac{1}{2}O_2(g) \rightarrow 4CO_2(g) + 5H_2O(l); \qquad \Delta H_{comb}^0$$
$$5H_2O(l) \rightarrow 5H_2(g) + 2\tfrac{1}{2}O_2(g); \qquad -5\Delta H_f^0[H_2O]$$
$$4CO_2(g) \rightarrow 4C(s) + 4O_2(g); \qquad -4\Delta H_f^0[CO_2]$$
$$C_4H_{10}(g) \rightarrow 4C(s) + 5H_2(g); \qquad -\Delta H_f^0[C_4H_{10}].$$

Therefore $-\Delta H_f^0[\text{n-}C_4H_{10}] = \Delta H_{comb}^0 - 5\Delta H_f^0[H_2O] - 4\Delta H_f^0[CO_2]$. At

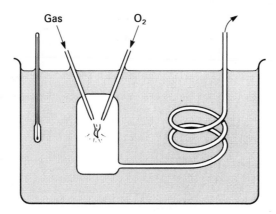

Fig. 5.5 Flame calorimeter.

298 K such a calculation gives

$$- \Delta H_f^0 [\text{n-C}_4\text{H}_{10}] = - 2877.13 + 5 \times 285.85 + 4 \times 393.51 \text{ kJ mol}^{-1}$$

and $\Delta H_f^0 [\text{n-C}_4\text{H}_{10}] = - 126.16 \text{ kJ mol}^{-1}$.

A similar calculation gives

$$\Delta H_f^0 [\text{i-C}_4\text{H}_{10}] = - 134 \cdot 53 \text{ kJ mol}^{-1}.$$

The enthalpy change accompanying one mole of the isomerization reaction n-C$_4$H$_{10}$ → i-C$_4$H$_{10}$ is given (Section 5.2) by

$$\Delta H^0 = \Delta H_f^0 [\text{products}] - \Delta H_f^0 [\text{reactants}]$$
$$\Delta H^0 = - 134 \cdot 53 + 126 \cdot 16 = - 8 \cdot 37 \text{ kJ mol}^{-1}.$$

The isobutane is energetically more stable as it has the lower enthalpy.

Absolute entropy of n-butane

To determine the entropy of n-butane at 298 K we must measure the heat capacity from within a few degrees of the absolute zero of temperature up to room temperature. This is done in an adiabatic vacuum calorimeter (Fig. 5.6) in which the sample under investigation is thermally insulated by locating it in an evacuated enclosure. A known quantity of electrical heat can be added and the rise in temperature measured with a platinum resistance thermometer. After a correction for the heat capacity of the vessel, the heat capacity of the compound it contains may be directly calculated from the equation

$$C_P = \left(\frac{\partial H}{\partial T} \right)_P = \left(\frac{\mathrm{d}q}{\mathrm{d}T} \right)_P \qquad \text{(Section 2.8)}.$$

In practice many refinements to the simple design illustrated are necessary if

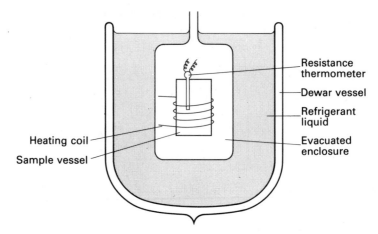

Fig. 5.6 Adiabatic vacuum calorimeter.

accurate results are required. The entropy is given by

$$S_T = \int_0^T \frac{C_P}{T} dT + \frac{\Delta H_{fus}}{T_{fus}} + \frac{\Delta H_{vap}}{T_{vap}}.$$

The calorimeter can be used to determine directly the enthalpy of fusion of the sample. The enthalpy of vaporization can either be measured directly in the apparatus or obtained more easily from vapour pressure measurements using the Clausius–Clapeyron equation (Sections 4.7 and 4.8),

$$\frac{d\ln P}{dT} = \frac{\Delta H_{vap}}{RT^2}.$$

The results of Aston and Messerly[†] for n-butane are illustrated in Fig. 5.3 and analysed in Table 5.3. An extra contribution from a solid-state phase transition at 107.55 K is included (and its magnitude emphasizes the importance of including all phase transitions). A similar experiment and calculation for isobutane gives $S_{298}^0 = 294.6 \, \text{J K}^{-1} \, \text{mol}^{-1}$. Thus

$$\Delta S^0 = 294.6 - 309.9 \, \text{J K}^{-1} \, \text{mol}^{-1}$$

$$= -15.3 \, \text{J K}^{-1} \, \text{mol}^{-1}.$$

Entropy considerations (unlike the energy) favour the n-butane. We can now obtain ΔG^0 at 298 K using the equation derived in Section 4.2.

$$\Delta G^0 = \Delta H^0 - T\Delta S^0$$

$$\Delta G^0 = -8.4 + 298 \times 15.3 \times 10^{-3} \, \text{kJ mol}^{-1}$$

Table 5.3 Entropy of n-butane

T/K	Method	$\Delta S/(\mathrm{J\,K^{-1}\,mol^{-1}})$
0–10	Extrapolation	0.62
10–107.55	$\int \dfrac{C_P}{T}$, Solid I	60.80
107.55	Transition, $\dfrac{\Delta H_{\mathrm{trans}}}{T_{\mathrm{trans}}}$	19.20
107.55–134.89	$\int \dfrac{C_P}{T}$, Solid II	18.91
134.89	Melting, $\dfrac{\Delta H_{\mathrm{fus}}}{T_{\mathrm{fus}}}$	34.54
134.89–272.66	$\int \dfrac{C_P}{T}$, Liquid	84.52
272.66	Vaporization, $\dfrac{\Delta H_{\mathrm{vap}}}{T_{\mathrm{vap}}}$	82.09
272.66–298.16	$\int \dfrac{C_P}{T}$, Gas	9.17
		$S^{0}_{298} = 309.9$

$\Delta H_{\mathrm{trans}} = 2067 \mathrm{\ J\,mol^{-1}}$
$\Delta H_{\mathrm{fus}} = 4660 \mathrm{\ J\,mol^{-1}}$
$\Delta H_{\mathrm{vap}} = 22.388 \mathrm{\ J\,mol^{-1}}$

$$\Delta G^0 = -8.4 + 4.6 \mathrm{\ kJ\,mol^{-1}}$$
$$\Delta G^0 = -3.8 \mathrm{\ kJ\,mol^{-1}}.$$

As the free-energy change is negative at this temperature we know that isobutane will predominate in the equilibrium mixture.

Determination of equilibrium constant

This value of ΔG^0 can be used to calculate the equilibrium constant for the isomerization reaction at 298 K;

$$\Delta G^0 = -RT \ln K_P \quad \text{(Section 4.11),} \quad \text{or} \quad \ln K_P = -\frac{\Delta G^0}{RT}$$

$$\lg K_P = \frac{-\Delta G^0}{2.3\,RT} = \frac{3800}{2.3 \times 8.3 \times 298} = 0.67$$

so $K_P = 4.6$.

To calculate the composition of the equilibrium mixture at 298 K we write

$$n\text{-}C_4H_{10} \rightleftharpoons i\text{-}C_4H_{10}$$

$$x_n \qquad\qquad x_i$$

where x_n and x_i are the mole fractions of n-butane and i-butane. Then

$$K_P = P_i/P_n = x_i/x_n = 4.6 \quad \text{and} \quad x_n + x_i = 1.$$

Solving these equations gives $x_n = 0.18$ and $x_i = 0.82$. The fact that the iso-form predominates in the equilibrium mixture has been verified by direct observation of the equilibrium at high pressures in the presence of a suitable catalyst, aluminium bromide.

Such a series of experiments as has been outlined above are rarely necessary nowadays. For most compounds the necessary experiments have already been carried out and the thermodynamic properties are tabulated as in Appendix 1. For instance, in the case of the butanes we would merely look up the standard free-energy changes on formation in thermodynamic tables. We find

$$\Delta G_f^0[n\text{-}C_4H_{10}] = -17.15 \text{ kJ mol}^{-1} \text{ at 298 K}$$

and

$$\Delta G_f^0[i\text{-}C_4H_{10}] = -20.92 \text{ kJ mol}^{-1} \text{ at 298 K.}$$

As

$$\Delta G^0 = \Delta G_f^0[\text{products}] - \Delta G_f^0[\text{reactants}] \qquad \text{(Section 5.5)},$$

$$\Delta G^0 = -3.77 \text{ kJ mol}^{-1}.$$

Hence we could calculate the equilibrium constant K_P for the isomerization reaction in a few minutes.

5.9 Calculation of thermodynamic quantities at temperatures other than 298 K

If we required the equilibrium constant at a temperature other than 298 K we could calculate it readily from thermodynamic quantities using the Van't Hoff Isochore (Section 4.12).

$$\frac{\mathrm{d}\ln K_P}{\mathrm{d}T} = \frac{\Delta H^0}{RT^2}.$$

In its integrated form (assuming ΔH^0 is independent of temperature)

$$\ln K_P = -\frac{\Delta H^0}{RT} + \text{const}, \quad \text{or}$$

$$\ln\left(\frac{K_2}{K_1}\right) = -\frac{\Delta H^0}{R}\left(\frac{1}{T_2} - \frac{1}{T_1}\right),$$

dropping the subscript P in K_P for convenience and writing K_1 and K_2 for the equilibrium constants at T_1 and T_2. Let us calculate the equilibrium constant at 800 K, knowing that $K = 4.6$ and $\Delta H^\circ = -8.4\,\text{kJ mol}^{-1}$ at 298 K.

$$\lg\left(\frac{K_2}{4.6}\right) = -\frac{\Delta H^0}{2.3 \times 8.31}\left(\frac{1}{800} - \frac{1}{298}\right)$$

$$= +\frac{8400}{19.11} \times (1.25 - 3.35) \times 10^{-3} = -0.923.$$

Therefore $K_2 = 0.12 \times 4.6 = 0.55$ and the mixture contains only 35% isobutane.

At high temperatures the equilibrium is displaced in the direction of higher concentrations of the n-butane because the unfavourable entropy of the reaction to form isobutane becomes relatively more important as the temperature is raised.

The assumption that ΔH^0 is not a function of temperature is not essential to the calculation. A knowledge of the heat capacities of the isomers would enable the temperature variation of ΔH^0 to be accounted for (see Section 5.4).

In the simplest case we can regard the heat capacities of the isomers as independent of temperature. Then

$$\Delta H_T^0 = \Delta H_{298}^0 + \Delta C_P(T - 298)$$

and the Van't Hoff Isochore becomes

$$\frac{d \ln K_P}{dT} = \frac{\Delta H_T^0}{RT^2} = \frac{(\Delta H_{298}^0 - 298\Delta C_P)}{RT^2} + \frac{\Delta C_P}{RT}.$$

Integrating we obtain

$$\ln\left(\frac{K_T}{K_{298}}\right) = -\frac{(\Delta H_{298}^0 - 298\Delta C_P)}{R}\left[\frac{1}{T} - \frac{1}{298}\right] + \frac{\Delta C_P}{R}\ln\left(\frac{T}{298}\right).$$

For the butane isomerization $\Delta H_{298}^0 = -8.4\,\text{kJ}$ and $\Delta C_P = 96.82 - 97.45 = -0.63\,\text{J K}^{-1}$ (see Appendix 1). Substituting these values into the above equation gives $K_{800} = 0.54$, almost identical with the value obtained using the Van't Hoff Isochore in its simplest form. However, the difference can be significant for those reactions for which ΔC_P is large.

5.10 Ellingham diagrams

Many metals are extracted by reduction of their oxides with carbon. Under these circumstances the carbon can be regarded as competing with the metal to bind oxygen:

$$MO + C = M + CO$$
$$MO + \tfrac{1}{2}C = M + \tfrac{1}{2}CO_2.$$

If the free energy accompanying the combination of oxygen with carbon is more negative than that accompanying the binding of oxygen to the metal then the reduction of the metal will proceed. Thus if the free energy of formation of one of the oxides of carbon is more negative than the free energy of formation of the metal oxide, the carbon will remove the oxygen from the metal oxide.

The magnitudes of the free energies of formation can be illustrated diagramatically (Fig. 5.7). Such a plot is often termed an Ellingham diagram. We note that it is conventionally plotted with the negative values of ΔG_f^0 at the top. This means that if the line for uppermost oxide of carbon lies above the line representing the metal oxide, reduction will occur. Then in moving from the metal oxide to the oxide of carbon the change in free energy will be negative. (Remember an upward movement on the diagram represents a negative change.)

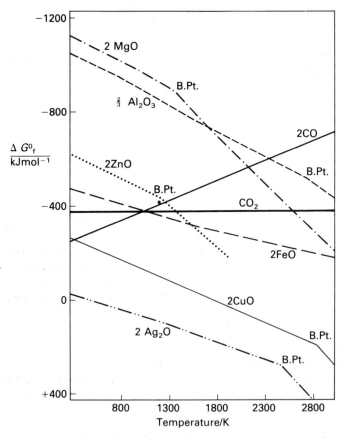

Fig. 5.7. The standard Gibbs free energy of formation of oxides as a function of temperature. This is called an Ellingham diagram.

The left-hand side of the diagram represents the standard free energies of formation at room temperature. The relative order tells us that only Ag_2O could be reduced by carbon under these conditions.

The capacity of carbon to reduce oxides varies with temperature since the free energies of formation vary with temperature as

$$\left(\frac{\partial \Delta G_f^0}{\partial T}\right)_P = - \Delta S_f^0$$

We find that whereas copper can be reduced at any temperature greater than room temperature, iron can only be extracted at temperatures above 1200 K. Zinc and magnesium can only be reduced at temperatures above 1300 K and 2100 K respectively, under which conditions they exist as vapours. Al_2O_3 is stable up to 2300 K.

We can also see from the diagram that above about 500 K Ag_2O is unstable and will dissociate into its elements since its free energy of formation becomes positive above this temperature. CuO is unstable above 1800 K.

Similar diagrams can be constructed for sulphides but carbon has only a low affinity for sulphur and is rarely able to reduce sulphide ores.

5.11 Free-energy functions

Calculations such as those described in Section 5.9 can become very laborious when ΔC_P must be expressed as a function of temperature. To make it easier to estimate ΔG^0 at temperatures other than 298 K it is convenient to tabulate functions that vary more slowly with temperature than ΔG_f^0. The function $(G^0 - H_0^0/T)$ (called rather misleadingly the free energy function) meets the requirements. For a reaction

$$\Delta G^0 = T\Delta\left(\frac{G^0 - H_0^0}{T}\right) + \Delta(H_0^0).$$

The second term, $\Delta(H_0^0)$, can be identified with the difference in the enthalpies of formation at absolute zero:

$$\Delta(H_0^0) = [\Delta H_{0f}^0]_{products} - [\Delta H_{0f}^0]_{reactants}.$$

ΔH_{0f}^0, usually written just ΔH_0^0, is included in tables of free-energy functions. The tables often include the function $(H^0 - H_0^0)/T$ from which ΔH^0 for a reaction can be calculated.

To illustrate the value of the method we will again calculate the equilibrium constant for the butane isomerization reaction at 800 K using the free-energy function data in Table 5.4.

Substituting in the equation

$$\Delta G^0 = T\Delta\left(\frac{G^0 - H_0^0}{T}\right) + \Delta(\Delta H_{0f}^0),$$

Table 5.4 The free energy function for the butane isomers

T/K Substance	298.16	400	600	800	1000	ΔH_0^0/kJ

T/K	$-\left(\dfrac{G^0 - H_0^0}{T}\right)\Big/ \text{J K}^{-1}\text{mol}^{-1}$					
Substance	**298.16**	**400**	**600**	**800**	**1000**	ΔH_0^0/kJ
n-C_4H_{10}	244.93	265.73	301.29	333.17	362.23	−99.04
iso-C_4H_{10}	234.64	254.64	288.49	319.89	348.86	− 105.86

we obtain

$$\Delta G^0 = [800(-319.89 + 333.17) + (-105.86 + 99.04)10^3] \text{ J mol}^{-1}$$
$$= (13.28 \times 800 \times 10^{-3} - 6.82) \text{ kJ mol}^{-1}$$
$$= 3.80 \text{ kJ mol}^{-1} \text{ at 800 K}\dagger$$

and

$$K_P = \exp\left(\frac{-\Delta G^0}{RT}\right) = 0.56.$$

This is in satisfactory agreement with the value obtained above. A more complete account of the use of free energy functions is given in Pitzer and Brewer, *Thermodynamics*,‡ pp. 166–9 and 669–70 and the published data tables are reviewed by Bett, Rowlinson and Saville in *Thermodynamics for chemical engineers*, Athlone Press (1976), pp. 412–15.

Problems

5.1. 4.40×10^{-4} kg of benzene, when burnt at 298 K in a bomb calorimeter of heat capacity 10.500 J K^{-1}, causes a rise in temperature of 1.75 K. Calculate the enthalpy change on combustion and the enthalpy of formation of 1 mol of benzene from its elements. The enthalpies of formation at 298 K of H_2O and CO_2 are − 286 and − 393.5 kJ mol^{-1}. The molecular weight of benzene is 78.

5.2. Using the bond energies of Table 5.1, estimate the standard heat of formation of n-butane from its elements in their standard states. The enthalpy of vaporization of graphite is + 717 kJ mol^{-1}.

5.3. At 298 K the enthalpy change on combustion of methanol is − 727 kJ mol^{-1}, that of hydrogen is − 286 kJ mol^{-1}, and that of graphite − 394 kJ mol^{-1}. Calculate the standard enthalpy of formation of methanol.

5.4. The heat capacity of liquid water is 75 J K^{-1} mol^{-1} whereas that of water vapour is 3.5 J K^{-1} mol^{-1}, at 373 K. The enthalpy change on vaporization at 373 K is 40.3 kJ mol^{-1}. Estimate the enthalpy change on vaporization at 423 K.

† Note: ΔG^0 at 298 K is approximately − 3.80 kJ mol^{-1} (see p. 76). At 800 K $\Delta G^0 = + 3.80$ kJ mol^{-1}. Do not be confused by this unfortunate coincidence!
‡ *See* Further Reading

5.5. Calculate the entropy of liquid mercury at its melting point, 234.1 K. The standard entropy of mercury (at 298.2 K) is 77.4 J K^{-1} mol^{-1} and its heat capacity is 2.87 J K^{-1} mol^{-1}. (Assume the heat capacity of mercury is constant in the temperature range 298.2 − 234.1 K.)

5.6. Using the thermodynamic tables given in Appendix 1 calculate the partial pressure of NO_2 in equilibrium with N_2O_4 in (i) a mixture at 1 atm and 298 K and (ii) a mixture at 1 atm and 318 K.

6
Ideal solutions

6.1 The ideal solution

A large number of chemical processes are carried out in solution and the application of chemical thermodynamics to solutions is an important part of the subject. Solutions can be gaseous, liquid, or solid. In this chapter we shall be concerned largely with solutions that are in the liquid state, for instance, mixtures of two liquids or the solution of a solid in a liquid. It is often convenient to refer to the substance which predominates in a solution as the *solvent* and to the minor constituent as the *solute*. In some solutions the components are *miscible* in all proportions. Thus ethanol and water will mix to form a homogeneous mixture whatever the relative quantities of ethanol and water. Other components will show limited mutual solubility. For example, only a limited amount of sodium chloride can be dissolved in water at any particular temperature. However much NaCl we add to a beaker of water the concentration of the salt will not exceed the value corresponding to a saturated solution. Some pairs of non-ionic substances, such as phenol and water also show limited mutual solubility.

The concept of an ideal solution is of great value in thermodynamics. We shall define it, for the moment, as a solution in which the total vapour pressure is given by Raoult's Law,

$$P = x_1 P_1^* + x_2 P_2^*$$

where x_1 is the mole fraction of component 1 ($x_1 = n_1/n$) and P_1^* is the vapour pressure of pure component 1. This behaviour is approximated by a number of liquid mixtures in which the two components are very similar, for example benzene–toluene mixtures, and is illustrated in Fig. 6.1. For each component we may write

$$P_i = x_i P_i^*$$

where P_i is the partial pressure of i in the vapour in equilibrium with the solution. If the vapour follows the perfect-gas law the chemical potential of component i in this phase may be expressed

$$\mu_i(g) = \mu_i^0(g) + RT \ln (P_i/\text{atm}) \qquad \text{(Section 4.10)}$$

where P_i is the numerical value of the partial pressure when expressed in atmospheres. At equilibrium the chemical potential of i in the vapour phase

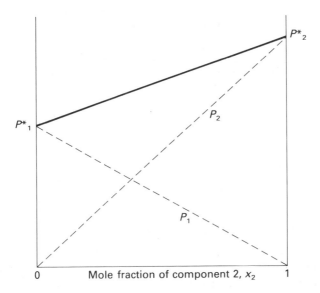

Fig. 6.1. Pressure of vapour in equilibrium with a liquid mixture that follows Raoult's Law. Broken lines are the partial vapour pressure of the components.

must be equal to the chemical potential of i in the solution. Therefore

$$\mu_i(\text{soln}) = \mu_i^0(\text{g}) + RT\ln(P_i/\text{atm}).$$

But as $P_i = x_i P_i^*$,

$$\mu_i(\text{soln}) = [\mu_i^0(\text{g}) + RT\ln(P_i^*/\text{atm})] + RT\ln x_i.$$

The term in the square brackets is constant at any temperature and is in fact the chemical potential of the pure liquid i, $\mu_i^*(\text{l})$, as for a pure liquid in equilibrium with its vapour

$$\mu_i^*(\text{l}) = \mu_i^0(\text{g}) + RT\ln(P_i^*/\text{atm}) = \mu_i(\text{g}).$$

The equation

$$\mu_i(\text{soln}) = \mu_i^*(\text{l}) + RT\ln x_i$$

holds for all components of an ideal solution and provides a more useful definition of ideal behaviour than one based on vapour pressures.

We must note when we write, for the perfect gas, the equation

$$\mu_i(\text{g}) = \mu_i^0(\text{g}) + RT\ln(P_i/\text{atm}),$$

that this equation is true for a perfect gas at all pressures. We have included the pressure variation of chemical potential. This is not true for the equation

$$\mu_i(\text{soln}) = \mu_i^0(\text{l}) + RT \ln x_i,$$

which is true only at 1 atm pressure. At other pressures we must include a term to allow for the fact that $\mu_i(\text{l})$ is a function of pressure, thus

$$\mu_i(\text{soln}) = \mu_i^0(\text{l}) + \left(\frac{\partial \mu_i}{\partial P}\right)_T \Delta P + RT \ln x_i.$$

As

$$\left(\frac{\partial \mu_i}{\partial P}\right)_T = \left[\frac{\partial}{\partial P}\left(\frac{\partial G}{\partial n_i}\right)\right]_T = V_i \quad \text{(Section 4.10)},$$

where V_i is the molar volume of i, we obtain

$$\mu_i(\text{soln}) = \mu_i^0(\text{l}) + V_i(\text{l})\Delta P + RT \ln x_i,$$

where ΔP is the excess pressure on the system (the total pressure less standard atmospheric pressure). As $\mu_i(\text{soln}) = \mu_i^* + RT \ln x_i$, the chemical potential in our new standard state μ_i^* (the pure substance at an arbitrary pressure) is related to μ_i^0 by

$$\mu_i^*(\text{l}) = \mu_i^0(\text{l}) + V_i(\text{l})\Delta P$$

and μ^* is pressure-dependent. Most pressures at which we perform experiments are close to 1 atm, and under these conditions $\mu^*(\text{l}) \approx \mu^0(\text{l})$.

The PV term is generally small and unless many atmospheres pressure are involved it is usually safe to ignore it (for instance when the vapour pressure of a liquid is not 1 atm). To simplify the notation for the standard states when applied to solutions we shall continue to use the standard states defined at 1 atm. We shall assume that the approximate equation

$$\mu_i(\text{soln}) = \mu_i^0(\text{l}) + RT \ln x_i$$

is a sufficiently accurate definition of an ideal solution for most purposes. In the few cases where pressures of many atmospheres are involved or where the effects of pressure are themselves being investigated, we shall apply an appropriate PV correction to μ.

Solutions whose components follow the equation

$$\mu(\text{soln}) = \mu^0(\text{l}) + RT \ln x_i$$

over all the composition range and at all temperatures of interest we shall call *truly ideal solutions*. We shall be concerned later with solutions whose components follow the equation only under restricted conditions.

The molecular basis for ideality is that the forces between the molecules of both components of the mixture should be identical. Just as the condition for a perfect gas is that there should be no intermolecular forces, so an ideal solution requires that all interactions, between like and unlike molecules, should be the same. Both the perfect gas and the ideal solution are limits to which the behaviour of real systems may approximate.

6.2 Properties of truly ideal solutions

Very few mixtures form truly ideal solutions in which both components follow the equation $\mu_i = \mu_i^0 + RT \ln x_i$ over all the composition range and over a range of temperatures. The total free energy of a solution is given by

$$G = \sum_i \mu_i n_i$$

This equation results from the integration of the relation $dG = \Sigma \mu_i dn_i$ (Section 4.10). For one mole of the solution, $G = \Sigma \mu_i(n_i/n)$ where $n = \Sigma n_i$ the total number of moles of substance in the solution, $n_i/n = x_i$ the mole fraction of component i, and $G = \Sigma \mu_i x_i$.

The change in free energy on mixing is the free energy of the solution less that of the isolated components.

$$\Delta G_{mix} = \Sigma \mu_i x_i - \Sigma \mu_i^0 x_i,$$

and as

$$\mu_i = \mu_i^0 + RT \ln x_i,$$

$\Delta G_{mix} = RT\Sigma x_i \ln x_i$ (again for a mole of solution).

The entropy of mixing can be calculated from the relation

$$\Delta S_{mix} = -\left(\frac{\partial \Delta G_{mix}}{\partial T}\right)_P \qquad \text{(Section 4.4)}$$

giving $\Delta S_{mix} = - R\Sigma x_i \ln x_i$.

The value of the entropy of mixing obtained using this equation is called the ideal entropy of mixing. It corresponds to completely random mixing of the components and represents the limiting behaviour not only of liquid mixtures but also of gas mixtures and of solid solutions. It is a natural consequence of the molecular conditions for ideality mentioned earlier that the molecules of all components must interact in an identical manner. If the molecules behave identically then we would expect any mixture of such molecules to be entirely random. The enthalpy change on mixing to form an ideal solution is given by

$$\Delta H_{mix} = - \Delta G_{mix} + T\Delta S_{mix} \qquad \text{(Section 4.2)}$$
$$= RT\sum_i x_i \ln x_i - RT\sum_i x_i \ln x_i = 0.$$

Thus there is no heat absorbed or released when components mix to form an ideal solution. Again this follows from the identical behaviour of the molecules comprising an ideal solution. The volume change on formation of an ideal solution is also zero (Section 4.3) as

$$\Delta V_{\mathrm{mix}} = \left(\frac{\partial \Delta G_{\mathrm{mix}}}{\partial P}\right)_T = \left(\frac{\partial RT\Sigma x_i \ln x_i}{\partial P}\right)_T = 0.$$

These properties of ideal solutions form a basis from which the behaviour of real solutions may depart to a greater or lesser extent, depending on the similarity of the component molecules.

6.3 Mixtures of liquids

Consider a mixture of two liquids A and B which form an ideal solution and for which the vapour pressures of A and B follow Raoult's Law. Such behaviour is illustrated in Fig. 6.2 where the upper solid line represents the vapour pressure of the solution as a function of its composition. Let us now consider the composition of the vapour in equilibrium with a liquid of composition $x_B(l)$.

$$x_B(g) = \frac{P_B}{P} = \frac{x_B(l)P_B^*}{P}, \quad \text{and} \quad x_A(g) = \frac{P_A}{P} = \frac{x_A(l)P_A^*}{P},$$

where P is the total pressure ($P = P_A + P_B$). From these equations we obtain

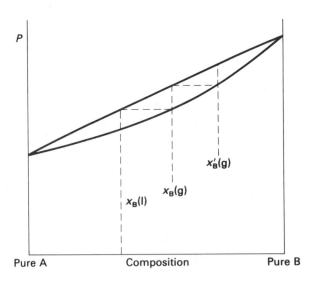

Fig. 6.2. A mixture of two liquids that follows Raoult's Law. The total vapour pressure of the mixture is plotted as a function of the composition of the liquid (upper line) and of the vapour (lower line). The upper line is sometimes referred to as the *bubble-point* line and the lower as the *dew-point* line.

the relation

$$\frac{x_A(g)}{x_B(g)} = \frac{x_A(l)}{x_B(l)} \cdot \frac{P_A^*}{P_B^*}.$$

This tells us that the vapour is richer in the more volatile component (B) than the liquid with which it is in equilibrium. Using this relation we can now construct another curve, the lower solid line in Fig. 6.2, which tells us the composition of the vapour in equilibrium with liquids of various compositions. It is this difference in the compositions of liquid and vapour that enables us to separate components of a mixture by distillation. If we take a solution of composition $x_B(1)$ and allow it to boil, the initial composition of the vapour will be $x_B(g)$, which is richer in component B. If the vapour is condensed and collected the process can be repeated and a liquid richer in component B, of composition $x_B'(g)$, may be collected. When the original liquid is boiled, since relatively more of component B is lost to the vapour, the remaining liquid becomes richer in component A.

We can regard a simple distillation as equivalent to one step on our diagram. Complex distillation columns carry out the equivalent of many steps on the diagram and their efficiency is defined in terms of the number of these steps—called in technical jargon the number of 'theoretical plates' of the column. The more plates the greater the separation of the components of the solution achieved in the distillation column.

Example

At 353 K the vapour pressures of benzene and bromobenzene are 757 and 66 mmHg respectively. For an equimolar mixture $[x_A(l) = x_B(l) = 0.5]$ calculate the total vapour pressure and the mole fraction of benzene in the vapour phases. Assume that the mixture follows Raoult's Law.

As the total pressure

$$P = P_A + P_B = x_A(l)P_A^* + x_B(l)P_B^*$$

$$= \tfrac{1}{2}(757 + 66) = \underline{411 \text{ mmHg}}$$

This is close to the experimental value of 407 mmHg showing that the solution follows Raoult's Law quite closely.

The mole fraction of benzene in the vapour phase may be estimated by

$$x_A(g) = x_A(l)P_A^*/P$$

$$= \frac{0.50 \times 757}{411} = 0.92.$$

The vapour is richer in the more volatile component.

6.4 Ideal solutions of solids in liquids

Consider a solid dissolving in a liquid to form an ideal solution (Fig. 6.3). The solute in its solid state will have the lower energy but the dissolution process is favoured by the gain in entropy as the solute becomes dispersed in the solution. At equilibrium the chemical potential of the solid must be equal to the chemical potential of the same substance in the solution. For an ideal solution (Section 6.1)

$$\mu_2^0(s) = \mu_2^0(\text{soln}) = \mu_2^0(l) + RT \ln x_2.$$

(The subscript 2 is usually used to denote the solute and subscript 1 the solvent.) Therefore

$$RT \ln x_2 = \mu_2^0(s) - \mu_2^0(l) = -\Delta G_{2\text{fus}}^0$$

where $\Delta G_{2\text{fus}}^0$ is the free-energy change on melting (fusion). From the Gibbs–Helmholtz equation (Section 4.4)

$$\left[\frac{\partial \left(\frac{\Delta G_2}{T} \right)}{\partial T} \right]_P = -\frac{\Delta H_2}{T^2}$$

$$\left(\frac{\partial \ln x_2}{\partial T} \right)_P = \frac{\Delta H_{\text{fus}}^0}{RT^2}.$$

ΔH_{fus}^0 is the heat of fusion of the solid. (We shall drop the subscript 2 from ΔH_{fus}^0.) At the melting point of the solid, T_{fus}, the solubility x_2 will be unity, since two liquids which form an ideal solution are completely miscible (they will mix in all proportions to form a homogeneous solution). We can integrate the equation above:

$$\ln x_2 - \ln(1) = \int_{T_{\text{fus}}}^{T} \frac{\Delta H_{\text{fus}}^0}{RT^2} \, dT = -\left[\frac{\Delta H_{\text{fus}}^0}{RT} \right]_{T_{\text{fus}}}^{T}$$

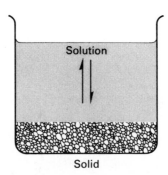

Solution

Solid

Fig. 6.3. Equilibrium between a solute in the solid state and in solution.

or

$$\ln x_2 = \frac{\Delta H_{fus}^0}{R} \left(\frac{1}{T_{fus}} - \frac{1}{T} \right).$$

Figure 6.4 shows that the ideal solubility equation for solids gives a good prediction of the solubility of naphthalene in benzene (circles) but not of its solubility in cyclohexane (squares).

Example

From a knowledge of the heat of fusion and the melting point of a solute we can calculate its ideal solubility. Thus naphthalene melts at 353.2 K and its heat of fusion is 19.0 kJ mol^{-1}. We can apply the ideal solubility equation to estimate its ideal solubility in a liquid (any liquid) at 298 K.

$$\ln x = \frac{\Delta H_{fus}}{R} \left(\frac{1}{T_{fus}} - \frac{1}{T} \right)$$

$$\lg x = \frac{19\,000}{2.3 \times 8.3} (0.00283 - 0.00336) = -0.53$$

$$x = 0.30.$$

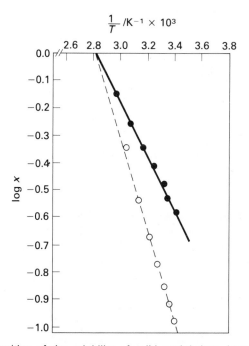

Fig. 6.4. The logarithm of the solubility of solid naphthalene (expressed as a mole fraction) as a function of reciprocal temperature. ● naphthalene in benzene, ○ naphthalene in cyclohexane. The solid line is calculated using the ideal-solubility equation.

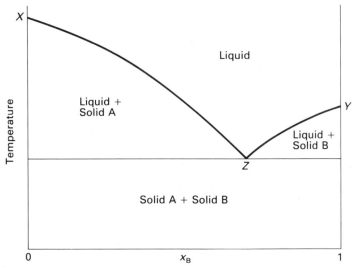

Fig. 6.5. A system showing a eutectic point.

In benzene the observed solubility of naphthalene at 298 K corresponds to a mole fraction of 0.29. However, the saturated solubility of this solute in hexane is given by $x = 0.12$ and in methanol by $x = 0.025$. The ideal solubility relation must be used with caution when attempting to estimate the actual solubility of solids in liquids.

Consider two mutually soluble substances A and B. The presence of a quantity of B will lower the freezing point of A and in the same way A will lower the freezing point of B. This is illustrated in Fig. 6.5.

The line of XZ can be regarded as either the freezing point curve of A in the presence of B or the solubility curve of B in A. Thus if the solid formed is solvent (the substance present in greater quantity), as for instance when ice freezes out of a salt solution, we usually call it a freezing point curve. Conversely when solid solute appears, as when a solution of iodine in benzene is cooled, we consider such lines solubility curves.

The two curves XZ and YZ meet at Z which is called the *eutectic point*. On cooling the mixture to this point the temperature remains constant until all the remaining liquid has frozen to produce a mixture of solids A and B. The coordinates of a eutectic point can be estimated by applying the ideal solubility equation to both components to determine the curves XZ and YZ.

6.5 Ideal dilute solutions

Though few solutions are truly ideal, essentially all show the characteristics of ideality when dilute. Fig. 6.6 illustrates that in a dilute solution the partial

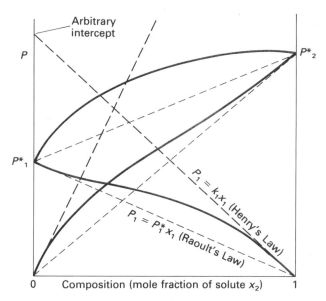

Fig. 6.6. The vapour pressures of the components and the total vapour pressure of a liquid mixture that deviates positively from Raoult's Law (schematic). Water–ethanol mixtures deviate from Raoult's Law in this manner.

pressure of the solvent follows Raoult's Law. Thus for the solvent, the substance in the highest concentration, we may write

$$\mu_i = \mu_i^0 + RT \ln x_i \qquad \text{(Section 6.1)}.$$

We also observe that when the solvent follows Raoult's Law the partial pressure of the solute follows an arbitrary straight line which in the example illustrated is not the Raoult's Law line. This line will intercept the $x_2 = 1$ axis not at P_2^* but at some arbitrary point. We will reserve discussion of the behaviour of the solute in such dilute solutions until later. However, even without considering the properties of the solute, the very fact that the solvent follows Raoult's Law is of enormous value to us. It enables us to treat the so-called colligative properties of solutions in a very straightforward manner.

6.6 Colligative properties

The colligative properties are those properties of liquid solutions which depend mainly on the number rather than the nature of the solute molecules present. The lowering of vapour pressure of a liquid by an involatile solid solute is a good example of such a property. To a large degree the lowering of vapour pressure depends not on the nature of the solute but only on the number of moles of solute present. If the vapour pressure of the solvent

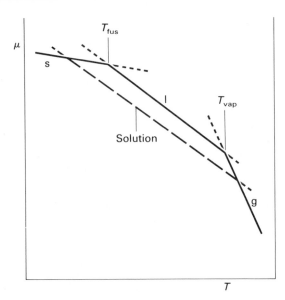

Fig. 6.7. Effect of added solute on the chemical potential of a liquid solvent as a function of temperature. The freezing point is depressed and the boiling point is elevated.

follows Raoult's Law then $P_1 = P_1^* x_1$. The solute enters into this equation only in so far as its mole fraction x_2 causes x_1, the mole fraction of the solvent, to be less than unity (as $x_1 + x_2 = 1$). If the solute is involatile and does not form solid solutions with the solvent when the latter is frozen, then the presence of the solute will lower the chemical potential of the solvent in the liquid phase of the system, as $\mu_1(\text{soln}) = \mu_1^0(\text{l}) + RT \ln x_1$, but will not affect the vapour or solid phases. This is illustrated in Fig. 6.7. The result of this lowering of the chemical potential of the liquid phase is to lower the freezing point and to elevate the boiling point of the solvent. The depression of freezing point and the elevation of boiling point are, like the depression of the vapour pressure itself, colligative properties.

6.7 Freezing-point depression

If a solute is dissolved in a liquid the freezing point of the liquid is usually lowered. The thermodynamics of this process are very simple if the solute dissolves only in the liquid phase of the solvent and does not form solid solutions with it. Then the phase that separates out on cooling is the pure solvent in its solid state.

Equating the chemical potentials of the *solvent* A in both phases,

$$A(s) \rightleftharpoons A(\text{soln})$$
$$\mu_1^0(s) = \mu_1(\text{soln}) = \mu_1^0(\text{l}) + RT \ln x_1 \qquad \text{(Section 6.1).}$$

If the subscript 1 (used to denote that it is the solvent that is in equilibrium) is dropped from all but the mole fraction term x_1, then

$$RT \ln x_1 = -\{\mu_1^0(l) - \mu_1^0(s)\} = -\Delta G_{fus}^0;$$

and differentiating with respect to T, using the Gibbs–Helmholtz relationship for the variation of $(\Delta G_{fus}^0/T)$ with temperature (Section 4.4), we obtain

$$\left(\frac{\partial \ln x_1}{\partial T}\right)_P = \frac{\Delta H_{fus}^0}{RT^2}.$$

We notice that this equation is similar to that obtained for the ideal solubility of solids (Section 6.4). Indeed the two situations are thermodynamically identical, but the enthalpy of fusion is that of the *solute* in the ideal solubility equation. In the case now being examined the solid state is the pure solvent (Fig. 6.8). We may integrate as before from $x_1 = 1$, $T = T_{fus}$ to $x_1 = x_1$, $T = T$ and obtain

$$\ln x_1 = \frac{\Delta H_{fus}^0}{R}\left(\frac{1}{T_{fus}} - \frac{1}{T}\right)$$

where T_{fus} is the freezing point of the pure solvent and T is its freezing point in the solution. This equation may be used to determine the molecular masses of solutes. A known mass of solute is added to a known quantity of solvent and the new freezing point recorded. The right-hand side of the equation can be evaluated if ΔH_{fus}^0 for the solvent is known, and hence x_2 can be determined.

Now

$$1 - x_1 = x_2 = \frac{n_2}{n_1 + n_2} = \frac{w_2/M_2}{w_1/M_1 + w_2/M_2},$$

where w_1 and w_2 are the masses of solvent and solute in the solution, and M_1

Pure solid solvent

Fig. 6.8. Equilibrium between a solvent in the pure solid state and in solution.

and M_2 their molecular masses (for definition see Section 1.5). If w_1 and w_2, and M_1, the molecular mass of the solvent, are known, then M_2, the molecular mass of the solute, may be evaluated.

Frequently the equation is used in a simplified form. The approximations commonly made are as follows:

$$\ln x_1 = \frac{\Delta H^0_{fus}}{R}\left(\frac{T - T_{fus}}{T_{fus}T}\right).$$

As the depression of freezing point is usually small we may write T^2_{fus} for $T_{fus}T$. Putting $\Delta T = T_{fus} - T$ we obtain

$$\ln x_1 = -\frac{\Delta H^0_{fus}}{R}\frac{\Delta T}{T^2_{fus}}.$$

For a dilute solution x_1 must be close to unity, so $\ln x_1 = \ln(1 - x_2) \approx -x_2$, and therefore

$$x_2 = +\frac{\Delta H^0_{fus}}{R}\frac{\Delta T}{T^2_{fus}}.$$

Furthermore for a dilute solution $n_1 \gg n_2$ and so

$$x_2 = \frac{n_2}{(n_1 + n_2)} \approx \frac{n_2}{n_1}.$$

Concentration is frequently expressed in terms of molality, m, which, though sometimes defined as a dimensionless quantity, we define for our present purposes as the number of moles of solute per kilogram of solvent, with units $mol\,kg^{-1}$.

$$m = \frac{n_2}{w_1} = \frac{w_2}{M_2 \times 10^{-3}\cdot w_1}\,mol\,kg^{-1}$$

where w_1 and w_2 are the masses of solvent and solute, and M_2 is the molecular mass of the solute. The factor 10^{-3} arises because molecular masses are expressed in $g\,mol^{-1}$ and must be converted to $kg\,mol^{-1}$ for the equations to be consistent.

Thus

$$m = \frac{n_2}{M_1 \times 10^{-3}\cdot n_1}\,mol\,kg^{-1}$$

and

$$x_2 = \frac{n_2}{n_1} = mM_1 \times 10^{-3}$$

where M_1 is the molecular mass of the solvent. Thus as $x_2 = mM_1/1000$, and we can write

$$\Delta T = K_{fus}m,$$

Table 6.1 Properties of common solvents

| Solvent | Boiling point | | $\Delta H_{fus}/$ (kJ mol^{-1}) | Melting point | $K_{vap}/$ K kg mol^{-1} | $K_{fus}/$ K kg mol^{-1} |
	$T_{vap}/$K	$\Delta H_{vap}/$ (kJ mol^{-1})		$T_{fus}/$K		
Water (H$_2$O)	373.2	40.7	6.01	273.2	0.512	1.86
Benzene (C$_6$H$_6$)	353.3	30.7	9.95	278.7	2.63	5.08
Ethanol (C$_2$H$_5$OH)	351.7	38.6	4.8	155.9	1.22	2.00
Carbon tetrachloride (CCl$_4$)	350.0	30.0	2.5	250.4	5.22	31.8
Acetic acid (CH$_3$CO$_2$H)	391.3	24.3	11.7	289.8	3.07	3.70

we obtain

$$K_{fus} = \frac{RT_{fus}^2 M_1}{\Delta H_{fus}^0 1000} \, \text{K kg mol}^{-1}.$$

Thus the molality, m, of a solution may be determined from the depression of freezing point. If the weight of solute, w, dissolved in 1 kg of solvent is known the molecular mass of the solute may be calculated from the relation $M_2 = 1000w/m$. K_{fus} is called the freezing-point-depression constant for the solvent. It may be readily calculated if the relevant properties of the solvent are known (Table 6.1), thus enabling the freezing-point equation to be used in a convenient form.

Example

The feeezing point of a solution that contains 18 g carbon tetrachloride (CCl_4) in 1000 g benzene is 0.60 K lower than that of pure benzene. Estimate the molecular mass of carbon tetrachloride. The freezing point constant for benzene, K_{fus}, is 5.08 K kg mol^{-1}.

$$\Delta T = K_{fus} \, m$$

therefore

$$m = \frac{\Delta T}{K_{fus}} = \frac{0.60}{5.08} = 0.12 \, \text{mol kg}^{-1}.$$

As

$$m = \frac{w_2 \, 1000}{M_2 \, w_1} \, \text{mol kg}^{-1} \qquad \text{(see above)}$$

$$M_2 = \frac{18 \times 1000}{0.12 \times 1000} = 150 \, \text{g mol}^{-1}.$$

This is a good estimate of the actual value 153.8 g mol^{-1}.

6.8 Elevation of boiling point

The addition of a non-volatile solute lowers the vapour pressure of a solvent and it will require a higher temperature before its vapour pressure becomes 1 atm. Thus the boiling point will be raised. If we consider the equilibrium

$$A(\text{soln}) \rightleftharpoons A(\text{g})$$

we have for the solvent

$$\mu_1(\text{soln}) = \mu_1^0(\text{l}) + RT \ln x_1 \qquad \text{(Section 6.1)}$$

and

$$\mu_1(\text{g}) = \mu_1^0(\text{g}) + RT \ln (P_1/\text{atm}) \qquad \text{(Section 4.10)}.$$

As the normal boiling point is when $P_1 = 1$ atm the second term, $RT \ln (P_1/\text{atm})$, is zero. Equating the chemical potentials (dropping subscript 1 for ΔG^0)

$$RT \ln x_1 = \Delta G_{vap}^0,$$

and differentiating, using the Gibbs–Helmholtz relationship (Section 4.4),

$$\left(\frac{\partial \ln x_1}{\partial T} \right)_P = - \frac{\Delta H_{vap}^0}{RT^2}.$$

Integrating between the limits set for the pure solvent $x_1 = 1$ when $T = T_{vap}$ to $x_1 = x_1$ when $T = T$,

$$\int_1^{x_1} d\ln x_1 = \int_{T_{vap}}^{T} - \frac{\Delta H_{vap}^0}{RT^2} dT,$$

we obtain

$$- \ln x_1 = \frac{\Delta H_{vap}^0}{R} \left[\frac{1}{T_{vap}} - \frac{1}{T} \right].$$

This equation may be used to determine molecular masses in exactly the same manner as the freezing-point equation. It may be simplified as before to give $\Delta T = K_{vap}m$ where m is the molality, ΔT is the elevation of the boiling point, and K_{vap} is the boiling-point-elevation constant:

$$K_{vap} = \frac{RT_{vap}^2 M_1}{\Delta H_{vap}^0 1000} \text{ K mol}^{-1} \text{kg}^{-1}.$$

Molecular masses in solution

When the state of a substance in solution is unknown the 'effective molecular mass' determined from colligative properties may provide valuable information.

Thus the association of substances leads to fewer molecules in solution and the observed molality, m_{obs}, calculated from colligative properties by equations such as

$$m = \frac{\Delta T}{K_{fus}} \qquad \text{(Section 6.7)}$$

will be lower than the formal molality based on the assumption that no association or dissociation occurs. If dissociation occurs the observed molality will be greater.

The observed molecular mass when association or dissociation occurs, M_{obs}, can be calculated by

$$M_{obs} = \frac{m}{m_{obs}} \cdot M = \frac{m M K_{fus}}{\Delta T}.$$

Here M is the monomer molecular mass and m the molality of the solution calculated assuming no association or dissociation. Similar equations apply to the elevation of boiling point.

Ethanoic (acetic) acid CH_3CO_2H has a molecular mass of 60. In benzene it may exhibit an apparent molecular mass of almost 120. This is because two acetic acid molecules can readily associate to form a dimer.

$$2CH_3CO_2H \rightleftharpoons (CH_3 \cdot CO_2H)_2$$
$$1 - \alpha \qquad\qquad \alpha/2$$

where α is the fraction of the acetic acid present as dimer. The effective molecular mass M_{obs} may be related to α and the molecular mass of the monomer, M.

M_{obs} = (molecular mass of dimer) × (relative number of dimer molecules)

+ (molecular mass of monomer)

× (relative number of monomer molecules).

$$M_{obs} = 2M \frac{(\alpha/2)}{(1 - \alpha/2)} + M \frac{(1 - \alpha)}{(1 - \alpha/2)},$$

$$\alpha = 2(M_{obs} - M)/M_{obs}.$$

The observed molecular mass can thus be used to determine the degree of association.

Example

The depression of the freezing point of benzene by ethanoic (acetic) acid is:

Concentration of ethanoic acid	Depression of freezing point
$0.00335 \, \text{mol kg}^{-1}$	$0.0156 \, \text{K}$
$0.0149 \, \text{mol kg}^{-1}$	$0.0539 \, \text{K}$
$0.2373 \, \text{mol kg}^{-1}$	$0.608 \, \text{K}$

What is the degree of association of ethanoic acid in these solutions?

Substituting the data for the first solution into the equation

$$M_{obs} = \frac{mM K_{fus}}{\Delta T}$$

and taking the value for K_{fus} for benzene as $5.08 \, \text{K kg mol}^{-1}$ and the molecular mass of monomeric ethanoic acid as $60.05 \, \text{g mol}^{-1}$, we obtain

$$M_{obs} = \frac{0.00335 \times 60.05 \times 5.08}{\Delta T} \, \text{g mol}^{-1}$$

$$= 65.5 \, \text{g mol}^{-1}.$$

This corresponds to a degree of association

$$\alpha = \frac{2(M_{\text{obs}} - M)}{M_{\text{obs}}} = \frac{2(65.5 - 60.05)}{65.5}$$

$$= 0.014.$$

For the other two solutions M_{obs} are 84.4 and 119.0 g mol^{-1} and the corresponding degrees of association are 0.58 and 0.99 respectively. In the most concentrated solution the ethanoic acid exists almost entirely in the form of dimers.

In a similar manner, if a substance dissociates

$$A_2 \rightleftharpoons A + A$$
$$(1 - \alpha) \quad \alpha \quad \alpha$$

the degree of *dissociation* can be calculated from the observed molecular mass

$$\alpha = \frac{M - M_{\text{obs}}}{M_{\text{obs}}}.$$

where M is the molecular mass of the undissociated substance.

6.9 Osmotic pressure

If a solution is separated from the pure solvent by a membrane which is permeable only to the solvent molecules, the net effect is for solvent molecules to diffuse into the solution as there is a greater concentration of these molecules in the pure solvent than in the solution. If a pressure is applied to the solution this effect can be counteracted and equilibrium restored. This pressure is equal to what is called the *osmotic pressure* of the solution, Π.

Let us consider the chemical potential of the solvent on both sides of the membrane, as illustrated in Fig. 6.9.

LHS $\mu_1^L(l) = \mu_1^0(l)$.

RHS $\mu_1^R(l) = \mu_1^0(l) + RT \ln x_1 + \left(\frac{\partial \mu_1}{\partial P}\right)_T \Delta P$ (Section 6.1).

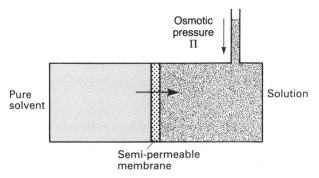

Fig. 6.9. Schematic representation of an osmotic pressure apparatus.

At equilibrium $\mu_1^L = \mu_1^R$; therefore $RT \ln x_1 = -\left(\frac{\partial \mu_1}{\partial P}\right)_T \Delta P.$

Now

$$\left(\frac{\partial \mu_1}{\partial P}\right)_T = \frac{\partial}{\partial n_1}\left(\frac{\partial G}{\partial P}\right)_T = V_1 \qquad \text{(Section 4.10)},$$

where V_1 is the molar volume of the solvent, and at equilibrium $\Delta P = \Pi$ the osmotic pressure, so $- \ln x_1 = V_1 \Pi / RT$. This equation may be simplified if we consider only very dilute solutions. Replacing x_1 by $(1 - x_2)$ and expanding the logarithm we get $x_2 = \Pi V_1 / RT$. For a dilute solution $x_2 \approx n_2/n_1$ and $V_1 = V/n_1$, where V is the total volume of the solution. Thus we obtain the approximate relation

$$\Pi = (RT/V)n_2,$$

or $\qquad\qquad\qquad\qquad \Pi = RTc$

where c is the concentration. If the osmotic pressure Π is in atmospheres and the concentration is in $mol\,dm^{-1}$ (equivalent to moles per litre in older units) the appropriate value of R is $0.086\,dm^3\,atm^{-1}\,K^{-1}\,mol^{-1}$.

Like other colligative properties, osmotic pressure measurements can be used to determine molecular masses.

Example

Calculate the osmotic pressure of a sucrose solution of concentration $0.10\,mol\,dm^{-3}$ at 303 K. The molecular mass of sucrose is $342.3\,g\,mol^{-1}$.

For this dilute solution we may assume that as $n_2 \ll n_1$, the mole fraction x_2 is given by

$$x_2 \approx \frac{n_2}{n_1}.$$

Then

$$\Pi = \frac{RT}{V}n_2 = RTc$$

$$\Pi = 0.082 \times 303 \times 0.10 = 2.48\,atm.$$

This is in excellent agreement with the experimentally observed value 2.47 atm.

6.10 Properties of the solute in dilute solutions

We have seen that in dilute solutions the solvent usually follows Raoult's Law whereas the solute does not. If we look at the molecular picture of a dilute

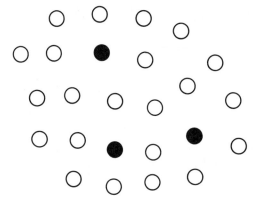

Fig. 6.10. Molecules of solvent ○ and solute ● in a dilute solution (schematic).

solution, as illustrated in Fig. 6.10, we can see why this should be so. The molecules of solvent are all surrounded by similar molecules apart from a few exceptions that have the other species as neighbours. Clearly, as the solution gets more dilute the limiting condition is for all solvent molecules to be in an environment of the same species, as occurs in the pure liquid. However, as the solution gets more dilute the limiting environment of the *solute* molecules is when each is surrounded entirely by solvent molecules. This clearly is not related to the state characteristic of pure liquid solute. Thus the solute's behaviour does not tend to that of the pure liquid solute and it does not follow Raoult's Law. Its partial vapour pressure may be a linear function of mole fraction in dilute solutions, but the constant of proportionality is not the vapour pressure of the pure solute P^* but merely an arbitrary constant, $P^\ominus$; thus

$$P_1 = x_1 P_1^* \text{ (solvent), and } P_2 = x_2 P^\ominus \text{ (solute) (see Fig. 6.6).}$$

If a solute has a partial vapour pressure proportional to its mole fraction it is said to follow Henry's Law. Raoult's Law may be regarded as a special case of Henry's Law. The chemical potential of a solute that follows Henry's Law is given by the expression

$$\mu_i = \mu_i^\ominus + RT \ln x_i$$

where $\mu_i^\ominus$ is an arbitrary constant, the chemical potential of a hypothetical liquid whose vapour pressure is $P^\ominus$. This equation can be of value even though we cannot determine $\mu_i^\ominus$ independently (as we can μ_i^0).

Consider a solute distributed between two immiscible liquid phases α and β. We have

$$\mu(\alpha) = \mu^\ominus(\alpha) + RT \ln x(\alpha)$$

and

$$\mu(\beta) = \mu^{\ominus}(\beta) + RT \ln x(\beta).$$

As $\mu^{\ominus}(\alpha)$ and $\mu^{\ominus}(\beta)$ are constant at any temperature and as $\mu(\alpha) = \mu(\beta)$ at equilibrium,

$$\frac{x(\alpha)}{x(\beta)} = \text{constant}.$$

Thus if the solute follows Henry's Law in both phases the ratio of its concentrations in the phases will be constant. This result is sometimes referred to as the *Nernst Distribution Law*.

6.11 Solubility of solids

We have already calculated the ideal solubility for a solid that follows Raoult's Law in solution and for which we can write

$$\mu_2(\text{soln}) = \mu_2^0(l) + RT \ln x_2 \qquad \text{(Section 6.1)}$$

where $\mu_2^0(l)$ is the chemical potential the solid would have if it were in the liquid state at the temperature of the experiment! This is not as ridiculous as it appears and $\mu_2^0(l)$ can be calculated for many solids without too much difficulty. However most solids in solution do not follow Raoult's Law, though in dilute solution they do adhere to Henry's Law. We may write for such solutes

$$\mu_2(\text{soln}) = \mu_2^{\ominus}(l) + RT \ln x_2 \qquad \text{(Section 6.10)},$$

and equating this with the chemical potential of the pure solid, for a saturated solution we obtain

$$RT \ln x_2 = \mu^0(s) - \mu^{\ominus}(l) = -\Delta G_{\text{mix}}.$$

Applying the Gibbs–Helmholtz equation (Section 4.4) as before we obtain

$$\left(\frac{\partial \ln x_2}{\partial T}\right)_P = \frac{\Delta H_{\text{mix}}}{RT^2}, \quad \text{and} \quad \ln x_2 = \frac{\Delta H_{\text{mix}}}{R}\left(\frac{1}{T_{\text{fus}}} - \frac{1}{T}\right).$$

This equation has the same form as that obtained for ideal solubility but ΔH_{fus}^0 has been replaced by the enthalpy of solution ΔH_{mix}. In non-ideal solutions of solids in liquids which do not follow either Henry's or Raoult's Laws, ΔH_{mix} is the *differential enthalpy of solution* of the solute in the saturated solution. Both ΔG_{mix} and ΔH_{mix} are for non-ideal solutions similar to the reaction free energy we introduced when studying equilibrium in chemical reactions. They are all differential quantities: ΔH_{mix} is the enthalpy change when one mole of solute is added to an infinite volume of nearly saturated solution, $\left(\frac{\partial H}{\partial n_2}\right)_{n_1}$. If we followed our previous practice we would write $\Delta H'_{mix}$ and $\Delta G'_{mix}$. It is important to distinguish the differential enthalpy of solution from the total or integral enthalpy of solution. This latter quantity is the total enthalpy change that would occur if one mole of solute were added to the appropriate quantity of solvent to make a saturated solution. The distinction may be more than merely academic in solutions that do not behave ideally. Thus when NaOH is dissolved in water, heat is evolved and the integral heat of solution is negative. However, the fact that the solubility of NaOH in water increases with rising temperature tells us that $\Delta H'_{mix}$ is positive.

Problems

6.1. At a given temperature the vapour pressures of two liquids which are completely miscible and form an ideal solution are 0.2 atm and 0.5 atm respectively. Estimate the mole fractions in both vapour and liquid phases at equilibrium when the total vapour pressure of the solution is 0.35 atm.

6.2. The vapour pressure of ether, $(C_2H_5)_2O$, is 445 mmHg at 293 K. That of a solution of 12.2×10^{-3} kg benzoic acid, $C_6H_5CO_2H$ in 0.100 kg ether is 413 mmHg. Calculate the molecular weight of the benzoic acid in ether.

6.3. A solution of 5×10^{-3} kg acetone, $(CH_3)_2CO$, in 1.000 kg of glacial acetic acid, CH_3CO_2H, froze at a temperature 0.32 K below the freezing point of the pure solvent. Calculate the freezing-point constant K_{fus} for glacial acetic acid.

6.4. 0.01 kg of naphthalene, $C_{10}H_8$, in 1.000 kg benzene, C_6H_6, lowers the freezing point 0.42 K from that of pure benzene, 278.8 K. Calculate the enthalpy of fusion of one mole of benzene.

6.5. A solution of 1.8×10^{-3} kg of a substance of high molecular weight in 1.00 kg of toluene, $C_6H_5CH_3$, has an osmotic pressure of 4.0 mm of toluene (density 860 kg m^{-3} at 298 K). Estimate the molecular weight of the substance.

6.6. Phenanthrene has an enthalpy of fusion of 18.6 kJ mol^{-1} and melts at 373 K. Calculate its ideal solubility in a liquid at 298 K expressed as a mole fraction.

7
Non-ideal solutions

7.1 The concept of activity

For many real solutions the equation $\mu_i = \mu_i^0(l) + RT \ln x_i$ (Section 6.1) does not hold for the solvent, and the expression which corresponds to Henry's Law (Section 6.10), $\mu_i = \mu_i^{\ominus}(l) + RT \ln x_i$ also fails to hold for the solute. In such circumstances it is convenient to introduce the concept of *activity*. We may write

$$\mu_i = \mu_i^0 + RT \ln a_i.$$

This equation defines the activity a_i of component i in terms of its chemical potential in the solution and in the standard state. This activity may be regarded as the *effective concentration* of component i relative to its standard state. (Activity is defined only in so far as the standard state to which it refers is also specified.) The effective concentration may, in a real solution, differ greatly from the true concentration as a result of the interactions between the molecules. The extent to which this is so is measured by the *activity coefficient* γ_i.

$$\gamma_i = \frac{a_i}{x_i} = \frac{\text{Effective concentration}}{\text{Real concentration}}.$$

This can be illustrated by consideration of a real liquid mixture.

Figure 7.1 shows the vapour pressure of a solution in which the vapour pressures of the components are said to deviate negatively from Raoult's Law (their vapour pressures are less than those predicted by the law). Writing the equations for the chemical potential of one component both in the solution and in the vapour phase,

$$\mu_i(g) = \mu_i^0(g) + RT \ln (P_i/\text{atm}) \qquad \text{(Section 4.10)},$$
$$\mu_i(\text{soln}) = \mu_i^0(l) + RT \ln a_i.$$

Now

$$\mu_i^0(l) = \mu_i^0(g) + RT \ln (P_i^*/\text{atm}) \qquad \text{(Section 6.1)},$$

as the pure liquid i must be in equilibrium with its vapour at the saturated

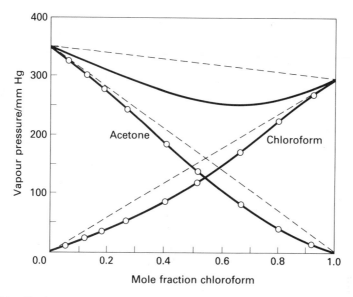

Fig. 7.1. Total vapour pressure and vapour pressures of components in a liquid mixture that deviates negatively from Raoult's Law.

vapour pressure. So we obtain, since $\mu_i(g) = \mu_i(\text{soln})$,

$$a_i = P_i/P_i^*.$$

This is in keeping with our definition of activity as the effective concentration. In the example we have taken $P_i < P_i(\text{ideal})$. The concentration of i in the vapour phase is less than would have been expected for a solution of such a concentration on the basis of ideal behaviour, and $a_i < x_i$. The effective concentration is less than the real concentration and the activity coefficient $\gamma_i(= a_i/x_i)$ is less than one. (In an ideal solution $a_i = x_i$ and $\gamma_i = 1$.) We can think of this result, $\gamma_i < 1$, as indicating that the component under investigation finds the solution a more 'congenial' environment than that of an ideal solution under the same conditions. Thus it shows less tendency to escape into the vapour phase.

Earlier we considered a system in which the components deviated positively from Raoult's Law, and had vapour pressures greater than the law would predict. In such a solution $a_i > x_i$, and $\gamma_i > 1$. For most normal liquid mixtures in which the components do not interact specifically (as in hydrogen-bond formation) this is the most common behaviour. The components show a greater tendency to escape into the vapour phase than in the corresponding ideal solution. In extreme cases the 'dislike' of the components for each other in the solution may cause the solution to separate into two phases at sufficiently low temperatures. The phenol–water system deviates

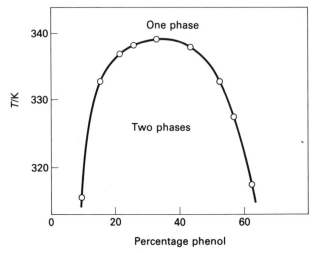

Fig. 7.2. Partial miscibility of phenol–water mixtures as a function of temperature.

strongly from Raoult's Law in a positive manner, and separates into two phases as illustrated in Fig. 7.2.

7.2 Activity of solids in liquids

When we looked at the solubility of naphthalene in various solvents (Section 6.4), we found that in benzene the actual solubility was close to the truly ideal value, as predicted on the basis of Raoult's Law, but in both hexane and methanol it was considerably lower. The chemical potential of the solid solute (and hence its activity in the solid state) is the same in all cases; the activity of the naphthalene in solution must also be identical, for at equilibrium

$$\mu(s) = \mu(\text{soln}) = \mu^0(l) + RT \ln a.$$

Nevertheless the concentrations are different, and therefore the activity coefficients must vary. In the ideal case

$$\mu(\text{soln}) = \mu^0(l) + RT \ln x_{id};$$

thus $a = x_{id}$ and $\gamma = 1$. For the solutions which deviate from ideality $\gamma = a/x = x_{id}/x$.

Thus for hexane $\gamma = 0.30/0.12 = 2.5$. The activity coefficient greater than unity indicates that a smaller quantity of naphthalene in solution is necessary for it to attain the chemical potential of solid naphthalene than if the solution were truly ideal. We could say that as the naphthalene 'does not like being in the solution' its tendency to stay in the solid state is greater.

We could have tackled the problem in another way. If we defined the activity of the solute naphthalene using the Henry's Law standard state $\mu_i^{\ominus}$ (from Section 6.10) we would obtain

$$\mu_i = \mu_i^{\ominus} + RT \ln a_i,$$

where a_i would *not* have the same value as that based on the pure-liquid standard state, which has chemical potential μ_i^0. Unlike μ_i^0, $\mu_i^{\ominus}$ would be different from system to system so that a different $\mu_i^{\ominus}$ value would be obtained for naphthalene in different solvents. The use of this standard state will be discussed in the next section.

As naphthalene in hexane follows Henry's Law (but not Raoult's Law) the activity coefficient defined on this basis would be approximately unity, as the equation $\mu_i = \mu_i^{\ominus} + RT \ln x_i$ holds for this system. Whether the pure-liquid or the Henry's Law standard state is used in any thermodynamic calculation is purely a matter of convenience. There is no fundamental thermodynamic objection to using *any* standard state. In the particular problems we are considering it is sometimes helpful to have the activity coefficient equal to unity so we can apply the equation $\mu_i = \mu_i^{\ominus} + RT \ln x_i$ to the system. On the other hand, if we wish to quantify the varying behaviour of naphthalene in different solvents the activity coefficients based on the pure-liquid standard state give us a convenient measure.

7.3 Activity in aqueous solutions

Many chemical experiments are carried out in aqueous solutions and it is important to be able to define activities in these circumstances. However, the standard state we have used so far—the pure liquid at one atmosphere pressure—is singularly inappropriate. We usually wish to express concentrations in molality (moles per kilogram of solvent) and for an electrolyte, such as sodium chloride, the pure-liquid state at room temperature is not a suitable reference state.

If the solute follows Henry's Law (Section 6.10), $\mu_i = \mu_i^{\ominus} + RT \ln x_i$. This is usually the case when the solution is very dilute and we may, as indicated in the previous section, define an activity such that

$$\mu_i = \mu_i^{\ominus} + RT \ln a_i = \mu_i^{\ominus} + RT \ln x_i \gamma_i,$$

and under these circumstances $a_i \to x_i$ as $x_i \to 0$. To refer the properties of the solute to its behaviour at infinite dilution clearly gives a more reasonable reference state. The actual reference state is in fact a solution for which the concentration term x_i is unity and whose properties are those of an infinitely dilute solution, for which $\gamma_i = 1$. This concept is more difficult to grasp than it is to use! We illustrate the standard state diagrammatically in Fig. 7.3. It is a pure liquid ($x_i = 1$) with a vapour pressure $P^{\ominus}$ and a chemical potential $\mu_i^{\ominus}$.

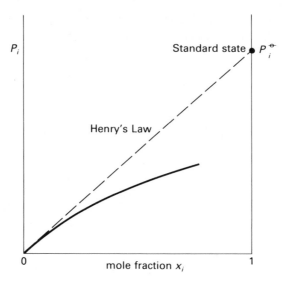

Fig. 7.3. Definition of a standard state for a solute, based on Henry's Law.

For an electrolyte solution the procedure is similar but we use unit molality, $m_i = 1 \text{ mol kg}^{-1}$, as the concentration of the standard state. Then for an ideal solution

$$\mu_i = \mu_i^{\ominus} + RT \ln \left(\frac{m_i}{\text{mol kg}^{-1}} \right)$$

and for a non-ideal solution

$$\mu_i = \mu_i^{\ominus} + RT \ln a_i$$

$$= \mu_i^{\ominus} + RT \ln \left[\frac{\gamma_i m_i}{\text{mol kg}^{-1}} \right].$$

Now $\mu_i^{\ominus}$ is the chemical potential of an ion in a (hypothetical) solution of unit molality which behaves like a solution of infinite dilution (where $\gamma_i = 1$, $m_i = 1 \text{ mol kg}^{-1}$, and $\mu_i = \mu_i^{\ominus}$).

Including the dimensions of molality makes the resulting equations clumsy and for this chapter we will adopt a simplifying convention that takes m to represent molality/mol kg^{-1}. Thus m now represents a *dimensionless ratio* equal to the number of moles of solute per kilogram of solvent.

The equation

$$\mu_i = \mu_i^{\ominus} + RT \ln a_i$$

must be modified when we apply the concept of activity to the electrolyte as a whole. Sodium chloride is essentially fully dissociated ($NaCl \rightarrow Na^+ + Cl^-$)

in aqueous solution. We can only deal with the overall activity a_{NaCl} as we have no way of determining a_{Na^+} and a_{Cl^-} independently. As each of the ions would be expected to follow Henry's Law independently, the total chemical potential can be written

$$\mu_{total} = \mu_{Na^+}^{\ominus} + \mu_{Cl^-}^{\ominus} + RT \ln a_{Na^+} + RT \ln a_{Cl^-}.$$

If we define the overall activity by the equation

$$\mu_{total} = \mu_{Na^+}^{\ominus} + \mu_{Cl^-}^{\ominus} + RT \ln a_{NaCl},$$

we can see by comparing these equations that

$$a_{NaCl} = a_{Na^+} \cdot a_{Cl^-}$$

and

$$a_{NaCl} = \gamma_{Na^+} m_{Na^+} \cdot \gamma_{Cl^-} m_{Cl^-}.$$

$m_{Na^+} = m_{Cl^-} = m$ the molality of the solution.
  We can define a mean ionic activity coefficient $\gamma_{\pm}$ by

$$\gamma_{\pm} = (\gamma_{Ag^+} \gamma_{Cl^-})^{1/2}.$$

Then

$$a_{NaCl} = (\gamma_{\pm} m)^2.$$

In dilute solutions, where $\gamma_{\pm} \to 1$, a_{NaCl} will be directly proportional to m^2. This is illustrated in Fig. 7.4. When the electrolyte is more complex the definition of overall activity must be modified accordingly:

$$\text{for} \quad M_p A_q \to pM^+ + qA^-, \qquad a_{M_p A_q} = (a_{M^+})^p \cdot (a_{A^-})^q$$

and

$$\gamma_{\pm}^{(p+q)} = (\gamma_{M^+})^p (\gamma_{A^-})^q.$$

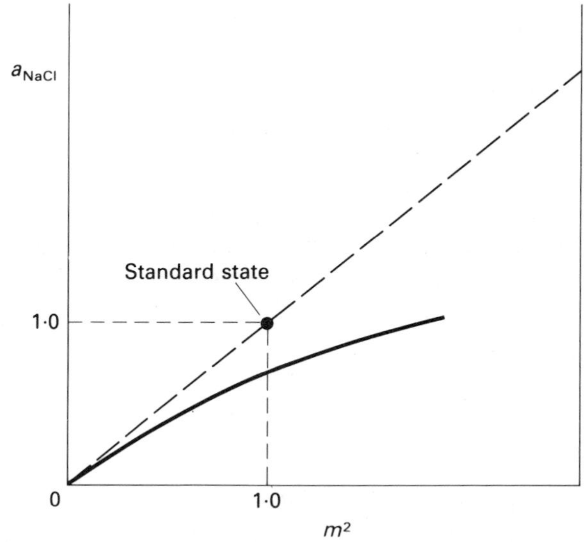

Fig. 7.4 Definition of a standard state for a 1:1 electrolyte based on Henry's Law.

Remember, both γ_i and a_i are different from those based on a pure-liquid standard state. The reference state has changed and a_i and γ_i are defined relative to that state. Strictly, in order to define the different activities a_i, we should write the three equations

$$\mu_i = \mu_i^0(x_i = 1) + RT \ln a_i, \quad \text{(see Section 7.1)}$$
$$\mu_i = \mu_i^\ominus(x_i = 1) + RT \ln a_i,$$

and

$$\mu_i = \mu_i^\ominus(m_i = 1) + RT \ln a_i.$$

In the examples that follow we shall always use the definition of activity based on unit molality as the standard state.

7.4 Chemical equilibria in solution

We have previously considered equilibria between perfect gases in which the chemical potential of each component followed the equation derived in Section 4.10:

$$\mu_i(g) = \mu_i^0(g) + RT \ln(P_i/\text{atm}).$$

In solution we may write for the chemical potential $\mu_i = \mu_i^\ominus + RT \ln a_i$ (Section 7.2), where the standard state is one of unit molality which behaves as if it were at infinite dilution. Let us consider the simplified equilibrium

$$A \rightleftharpoons B$$

$$m_A \quad m_B$$

$$\mu_A = \mu_A^\ominus + RT \ln a_A = \mu_A^\ominus + RT \ln \gamma_A m_A,$$

$$\mu_B = \mu_B^\ominus + RT \ln a_B = \mu_B^\ominus + RT \ln \gamma_B m_B.$$

At equilibrium $\mu_A = \mu_B$, and $\Delta G^\ominus = \mu_B^\ominus - \mu_A^\ominus = -RT \ln a_A/a_B = -RT \ln K_a$.

$\Delta G^\ominus$ is the change in free energy when a mole of A at unit molality is transformed into a mole of B at the same concentration, both solutions behaving as if they were at infinite dilution. In other words the free-energy change for one mole of reaction with both reactants in their standard states is $\Delta G^\ominus - RT \ln K_a$. K_a, the equilibrium constant expressed in terms of activities, is *exactly* constant at any given temperature and pressure. As $a_i = \gamma_i m_i$

$$K_a = \frac{\gamma_B m_B}{\gamma_A m_A} = \frac{\gamma_B}{\gamma_A} K_m$$

where K_m is the equilibrium constant expressed in terms of molalities. This is *not* an exact constant and is equal to K_a only when $\gamma_A = \gamma_B = 1$, that is when both reactants and products follow Henry's Law. Again it must be remembered that $\Delta G^{\ominus}$ and K_a, like a_i and γ_i, depend on the choice of standard state.

For solutions, as for gases, we may calculate the temperature dependence of K_a using the Van't Hoff Isochore (Section 4.12),

$$\left[\frac{\partial \ln K_a}{\partial T}\right]_P = \frac{\Delta H^{\ominus}}{RT^2}.$$

$\Delta H^{\ominus}$ is the difference in enthalpy between the products and reactants in their standard states for one mole of reaction. For systems not at equilibrium the change in free energy for a mole of reaction, all concentrations being maintained constant is, as before,

$$\Delta G' = \Delta G^{\ominus} + RT \ln a_B/a_A.$$

$\Delta G'$, the so-called reaction free energy, is $\left(\dfrac{\partial G}{\partial \xi}\right)_{T,P}$ where ξ is the extent of reaction (Section 4.11).

An important case of equilibrium in ionic solutions is the solubility of a sparingly soluble salt, for example, silver chloride

$$AgCl(s) \rightleftharpoons Ag^+(aq) + Cl^-(aq).$$

The chemical potential of the solid is constant, thus we obtain

$$\mu_{AgCl}(s) = \mu_{Ag^+}^{\ominus} + RT \ln a_{Ag^+} + \mu_{Cl^-}^{\ominus} + RT \ln a_{Cl^-}$$

and

$$RT(\ln a_{Ag^+} + \ln a_{Cl^-}) = \text{const}$$

or

$$a_{Ag^+} \cdot a_{Cl^-} = K_s.$$

The constant K_s is known as the solubility product of the salt:

$$K_s = (\gamma_{Ag^+} \gamma_{Cl^-}) m_{Ag^+} m_{Cl^-}.$$

If the salt is only slightly soluble the activity coefficients will be close to unity, thus

$$K_s = m_{Ag^+} m_{Cl^-} = m^2.$$

However, if we dissolve the salt in an electrolyte solution (with no ion in common) we find that the activity coefficients are usually less than unity and as $K_s = (\gamma_{Ag^+} \gamma_{Cl^-}) m^2$ is a constant, m^2 is greater than its value at infinite dilution. In other words the presence of the other ions in solution tends to *increase* the solubility of the sparingly soluble salt.

Thus we find that the solubility of thallium (I) chloride TlCl is 0.016 mol kg^{-1} in distilled water and 0.023 mol kg^{-1} in a 0.30 mol kg^{-1} solution of KNO_3. It can be calculated that the mean activity coefficient of the salt

$$\gamma_\pm = (\gamma_{Tl^+} \gamma_{Cl^-})^{\frac{1}{2}}$$

dissolved in pure water is 0.88. When TlCl is dissolved in $0.30 \, mol \, kg^{-1}$ KNO_3 the mean activity coefficient is 0.61.

We note that the ionic concentration produced by TlCl when it dissolves in water is sufficient to cause the activity coefficient to depart significantly from unity, its value at infinite dilution. A famous theory due to Debye and Hückel provides the explanation for this interesting behaviour.†

7.5   Electrochemical cells

The systems we have studied so far have not been harnessed to do work and have only done work by expanding against atmospheric pressure. Under these conditions the position of equilibrium is defined by the equation derived in Section 4.2:

$$dG = 0 \quad (T, P \, const).$$

However, the electrochemical cell is one important chemical system in which work other than PV work is performed. In this situation $dG = dw_{additional}$ is the appropriate equilibrium condition. In such a cell when the external circuit is open the electrodes will have different electrical potentials. These will arise from the tendency of the electrode material to give up electrons and to pass into solution as ions (or vice versa). On connecting the external circuit, chemical reactions take place which lead to electrons being removed from one electrode and transferred to the other. If an electric current is allowed to flow through an external circuit, linking the electrodes down the potential gradient, it can be used to do work.

A simple cell is illustrated in Fig. 7.5. One electrode is made of zinc surrounded by a solution of a zinc salt; the other is made of copper immersed in a solution of a copper salt. The solutions can be linked in such a way that ions can flow from one compartment to the other, but the connection, a 'salt bridge', is arranged so as to prevent the solutions mixing too rapidly. The tendency in such a cell is for the zinc to form ions and to give up electrons making the zinc electrode negative. Conversely, the copper ions in solution tend to gain electrons and be deposited as metallic copper. The overall reaction is

$$Zn + Cu^{++} \rightarrow Cu + Zn^{++}.$$

In the *external* circuit the flow of electrons is to the positive copper electrode from the negative zinc.

† See Atkins *Physical chemistry, fourth edition*, (Chapter 10).

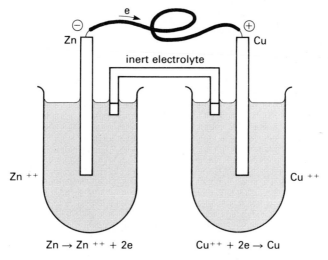

Fig. 7.5. A simple electrochemical cell.

The maximum electrical work done by such a cell is equal to the product of the charge flowing and the potential difference across which it flows. The work done *on* the cell is

$$w = -EQ \quad \text{(see Section 2.1)},$$

where E is the electromotive force of the cell and Q the charge flowing. For one mole of the cell reaction

$$Q = nN_A e$$

where N_A is Avogadro's number, $-e$ is the charge on the electron and n is the number of moles of electrons transferred per mole of reaction. The product $N_A e$, the magnitude of the total charge of a mole of electrons, is called the Faraday and written F. Thus

$$w = -nFE.$$

For the reaction we have considered, two electrons are transferred from the zinc to the copper for a mole of reaction and so $n = 2$.

If the cell is worked under reversible conditions (that is, the current is drawn infinitely slowly†) then we may regard the system as in equilibrium, and accordingly

$$dG = dw_{\text{additional}}(T, P \text{ const}) \quad \text{(Section 4.2)}.$$

For one mole of the cell reaction under constant conditions we may write

$$\Delta G' = -nFE$$

---

† To do this an external e.m.f. must be applied to balance the e.m.f. of the cell.

where $\Delta G'$ is the reaction free energy. (If the reaction were to proceed to a significant extent the concentrations of the electrolytes would change. As before $\Delta G'$ is therefore strictly the differential term $\left(\dfrac{\partial G}{\partial \xi}\right)_{T,P}$ where ξ is the extent of the reaction. See Section 4.11.)

As we can write for a simplified cell reaction (see Section 7.4),

$$A \rightleftharpoons B$$

$$\Delta G' = \Delta G^{\ominus} + RT\ln(a_B/a_A).$$

so

$$-nFE = -nFE^{\ominus} + RT\ln(a_B/a_A),$$

or

$$E = E^{\ominus} - (RT/nF)\ln(a_B/a_A).$$

For a more general cell reaction

$$aA + bB \rightleftharpoons nN + mM$$

we would obtain

$$E = E^{\ominus} - (RT/nF)\ln(a_N^n a_M^m/a_A^a a_B^b).$$

This equation is the *Nernst equation* which relates the e.m.f. E of a cell to $E^{\ominus}$, the e.m.f. of the cell when all the reactants and products are in their standard states. $E^{\ominus}$ is called the *standard e.m.f.* of the cell.

If the composition of the electrolyte solutions is such that the cell reaction is at equilibrium, there is no tendency for current to flow and the e.m.f. of the cell will be zero. Then

$$E = 0 \quad \text{and} \quad E^{\ominus} = \frac{RT}{nF}\ln K_a,$$

where K_a is the equilibrium constant of the cell reaction. The changes in other thermodynamic quantities can be obtained from the equation $\Delta G' = -nFE$.

As

$$-\Delta S' = \left[\frac{\partial(\Delta G')}{\partial T}\right]_P \qquad \text{(Section 4.4)},$$

$$\Delta S' = \left(\frac{\partial E}{\partial T}\right)_P nF,$$

and

$$\Delta H' = \Delta G' + T\Delta S' = -nFE + \left(\frac{\partial E}{\partial T}\right)_P nFT.$$

The primes indicate that the $\Delta H'$ and $\Delta S'$ refer to a mole of reaction carried out with the substances participating in the cell reaction maintained at a specified concentration. $\Delta H'$ is equal to the heat absorbed at constant pressure $(q)_P$ only if no work, other than PV work, is done. In an electrochemical cell work is done and the heat of reaction is *not* $\Delta H'$. We may calculate $(q)_P$ from the relation $\Delta S' = q_{rev}/T$ which gives the reversible heat absorption as $(q)_P = T\Delta S'$.

The electrochemical cell provides a very powerful way of determining equilibrium constants and the changes in thermodynamic properties accompanying reactions in solution.†

Example

The e.m.f. of the cell

$$Ag|AgCl(s)|HCl(aq)|Hg_2Cl_2|Hg$$

is 0.0421 V at 288 K and 0.0489 V at 308 K. Use this information to estimate ΔG, ΔS, and ΔH for the cell reaction

$$2Ag + Hg_2Cl_2 \rightarrow 2AgCl + 2Hg \text{ at } 298 \text{ K.}$$

The value of the e.m.f. at 298 K is taken as the mean of the values given, that is 0.0455 V. Then

$$\Delta G = -nFE = -2 \times 96\,500 \times 0.0455 \text{ J mol}^{-1}$$
$$= -8.78 \text{ kJ mol}^{-1}$$

$$\Delta S = nF\left(\frac{dE}{dT}\right) = 2 \times 96\,500\left(\frac{0.0489 - 0.0421}{20}\right) \text{J K}^{-1}\text{mol}^{-1}$$
$$= 65.6 \text{ J K}^{-1} \text{mol}^{-1}$$

$$\Delta H = \Delta G + T\Delta S = 10.76 \text{ kJ mol}^{-1}.$$

The negative value of ΔG results from the favourable entropy change that accompanies the reaction.

7.6 Standard electrode potentials

If we could measure the tendency of each possible half of an electrochemical cell (each electrode and the solution in which it is immersed) to take up or release electrons, then we could calculate the e.m.f. of a cell made up of two such electrodes. We cannot measure this tendency on an absolute basis but it

† The general background to electrochemistry will be found in Atkins *Physical chemistry*, *fourth edition* (Chapter 10) and the way electrochemical properties affect the inorganic properties of ions is described in G. Pass's *Ions in solution (3): Inorganic properties* (OCS7).

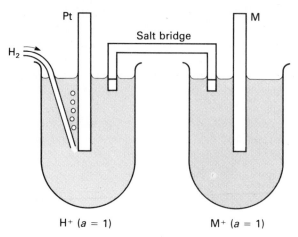

Fig. 7.6. Electrochemical cell with a hydrogen electrode set up to determine the *standard electrode potential* of element M.

is sufficient if we can measure it relative to some reference electrode. The standard chosen is the *standard hydrogen electrode* which is assigned the value $E^{\ominus} = 0$. It is made by bringing hydrogen gas at 1 atm into contact with a solution of its ions at unit activity at the surface of a platinum electrode.

To determine the *standard electrode potential* of an element M we set up a cell as illustrated in Fig. 7.6. The element is placed in a solution of its ions at unit activity (standard state, based on the unit-molality definition) and coupled to a standard hydrogen electrode.† The potential of element M with respect to the platinum of the hydrogen electrode is called the standard electrode potential of M. (If the element M is positive with respect to the hydrogen electrode then the standard electrode potential of M is positive and vice versa.) If the metal in the cell is zinc we find

$$E^{\ominus}(Zn^{++}, Zn) = -0.76 \text{ V.}$$

For copper

$$E^{\ominus}(Cu^{++}, Cu) = +0.34 \text{ V.}$$

As a zinc electrode tends to dissolve, forming zinc ions, the electrode tends to become negative, and in effect withdraws electrons from the solution:

$$Zn \rightarrow Zn^{++} + 2e^{-}.$$

On the other hand the copper electrode tends to lose electrons, becoming

---

† In practice we do not try to make up solutions of unit activity—this would require a knowledge of the activity coefficients of the ions in solution. Instead we can use very dilute solutions in the cell so that the activity coefficients are essentially unity. $E^{\ominus}$ can then be calculated using the Nernst equation (Section 7.5), replacing the activities by the concentrations of the ions.

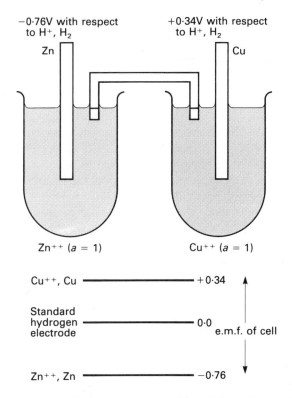

Fig. 7.7. Determination of the standard e.m.f. of a cell from the *standard electrode potentials* of the component half-cells.

positive. At this electrode the reaction must be

$$Cu^{++} + 2e^- \rightarrow Cu.$$

The e.m.f. of a cell with zinc and copper electrodes as in Fig. 7.7 can be calculated from a knowledge of the standard electrode potentials of these elements. The standard e.m.f. of the cell is the difference between the standard electrode potentials:

$$E^\ominus = E^\ominus_{r.h.s.} - E^\ominus_{l.h.s.} = 0.34 - (-0.76) = +1.10 \text{ V}.$$

We conventionally record the e.m.f.s of cells as the potential difference between the right and the left electrodes. Thus if we had reversed the cell, so that Zn was on the right, the sign of $E^\ominus$ would be reversed.

The standard electrode potential of an electrode is a measure of its tendency to gain electrons. If $E^\ominus$ is positive the electrode will tend to gain electrons; if $E^\ominus$ is negative the electrode will tend to lose electrons. Both these tendencies are measured relative to hydrogen. The alkali metals have the most negative electrode potentials whereas the halogen electrodes are very positive:

$$E^{\ominus}(Cl_2, Cl^-) = +1.36 \text{ V}.$$

We may look at this from a thermodynamic standpoint. Consider the electrode process $M^+ + e^- \rightarrow M$. If $E^{\ominus}$ for this electrode is negative, then as $\Delta G = -nFE$, the free-energy change will be positive. Thus the free-energy change for the reverse reaction $(M \rightarrow M^+ + e^-)$ will be negative and the cell reaction will go in this direction: this is consistent with the fact that the electrode becomes negative with respect to the hydrogen electrode.

The standard electrode potentials might well be called standard reduction potentials because they measure the tendency of the electrode material to be reduced by the gain of electrons. They naturally determine many of the properties of a substance. Because zinc has a more negative standard electrode potential than copper, elemental zinc will tend to reduce copper salts:

$$Zn + Cu^{++} \rightarrow Zn^{++} + Cu.$$

In other words zinc could be used to precipitate metallic copper from a

Table 7.1 Standard electrode potentials

Electrode	Process	E /V
Li^+, Li	$Li^+ + e \rightleftharpoons Li$	-3.045
K^+, K	$K^+ + e \rightleftharpoons K$	-2.925
Na^+, Na	$Na^+ + e \rightleftharpoons Na$	-2.714
Mg^{++}, Mg	$Mg^{++} + 2e \rightleftharpoons Mg$	-2.37
Zn^{++}, Zn	$Zn^{++} + 2e \rightleftharpoons Zn$	-0.763
Fe^{++}, Fe	$Fe^{++} + 2e \rightleftharpoons Fe$	-0.44
Sn^{++}, Sn	$Sn^{++} + 2e \rightleftharpoons Sn$	-0.136
Pb^{++}, Pb	$Pb^{++} + 2e \rightleftharpoons Pb$	-0.126
Fe^{3+}, Fe	$Fe^{3+} + 3e \rightleftharpoons Fe$	-0.036
H^+, H_2	$2H^+ + 2e \rightleftharpoons H_2$	[0]
Sn^{4+}, Sn^{++}	$Sn^{4+} + 2e \rightleftharpoons Sn^{++}$	0.15
Cu^{++}, Cu^+	$Cu^{++} + e \rightleftharpoons Cu^+$	0.153
Cu^{++}, Cu	$Cu^{++} + 2e \rightleftharpoons Cu$	0.34
I_2/I^-	$I_2 + 2e \rightleftharpoons 2I^-$	0.536
Fe^{3+}, Fe^{2+}	$Fe^{3+} + e \rightleftharpoons Fe^{2+}$	0.771
Ag^+, Ag	$Ag^+ + e \rightleftharpoons Ag$	0.799
Hg_2^{++}, Hg	$Hg_2^{++} + 2e \rightleftharpoons 2Hg$	0.799
Hg^{++}, Hg	$Hg^{++} + 2e \rightleftharpoons Hg$	0.854
Br_2, Br^-	$Br_2 + 2e \rightleftharpoons 2Br^-$	1.0652
O_2, OH^-	$O_2 + 4H^+ + 4e \rightleftharpoons 2H_2O$	1.229
Cl_2, Cl^-	$Cl_2 + 2e \rightleftharpoons 2Cl^-$	1.3595

copper sulphate solution. Those elements with negative electrode potentials are, in a similar way, able to displace hydrogen from acids:

$$Zn + 2HCl \rightarrow H_2 + ZnCl_2$$

(unless they become passive like aluminium).

We can apply the Nernst equation† to a half-cell, for example a metal in contact with a solution containing its ions

$$E = E^\ominus - \left(\frac{RT}{nF}\right) \ln\left(\frac{a_R}{a_0}\right)$$

where a_0 and a_R are the activities of the oxidized and reduced states. In the case of metals such as Zn and Cu one of the states is simply the electrode material itself. This need not be so. If a solution containing both Fe^{2+} and Fe^{3+} ions is brought into contact with an unreactive platinum electrode, an e.m.f. is set up owing to the reaction

$$Fe^{3+} + e^- \rightleftharpoons Fe^{2+}.$$

The Nernst equation may be applied to this electrode (relative to a standard hydrogen electrode) as before:

$$E = E^\ominus - \frac{RT}{F} \ln\left(\frac{a_{Fe^{2+}}}{a_{Fe^{3+}}}\right).$$

In this case $E^\ominus$ is found to be $+ 0.77$ V.

Problems

7.1. The vapour pressures of n-propanol and water at 298 K are 21.8 and 23.8 mm Hg respectively. In a solution in which the mole fraction of water is 0.20 their partial pressures are 17.8 and 13.4 mm Hg respectively. Calculate the activities and activity coefficients of the two components in this solution.

7.2. The standard electrode potential of zinc, $E^\ominus(Zn^{++}, Zn)$, is -0.76 V, and that of copper, $E^\ominus(Cu^{++}, Cu)$, is $+0.34$ V. Estimate the equilibrium constant at 298 K for the reaction $Cu^{++} + Zn \rightleftharpoons Cu + Zn^{++}$.

7.3. At 298 K the standard electrode potential of silver, $E^\ominus(Ag^+, Ag)$, is 0.80 V and the standard electrode potential of the cell $Pt(H_2)|HCl|AgCl|Ag$ is 0.22 V. Calculate the solubility product of AgCl at this temperature. (The electrode reaction at the AgCl, Ag electrode is $AgCl(s) + e \rightarrow Ag + Cl^-$.)

7.4. An electrochemical cell in which the reaction

$$Zn(s) + 2AgCl(s) \rightarrow ZnCl_2(aq) + 2Ag(s)$$

occurs has a standard e.m.f. of 1.055 V at 298 K and 1.015 V at 273 K. Estimate the changes in Gibbs free energy and the enthalpy accompanying one mole of the cell reaction at 298 K.

† See. p. 107, and P. W. Atkins (1990), p. 260.

8
Thermodynamics of gases

In following the most direct path from the principles of thermodynamics to the understanding of equilibrium in chemical systems, we have bypassed many useful thermodynamic relations that involve the properties of perfect and imperfect gases. These are summarized in this chapter.

8.1 Expansion of a perfect gas

We have already calculated, in Section 3.5, the changes in thermodynamic properties accompanying the *isothermal* expansion of a perfect gas. As $\left(\dfrac{\partial U}{\partial V}\right)_T = 0$ and $dU = \left(\dfrac{\partial U}{\partial T}\right)_V dT + \left(\dfrac{\partial U}{\partial V}\right)_T dV$ we have for an isothermal expansion $dU = 0$ and $dq = -dw = +PdV$, therefore

$$q = -w = \int_{V_1}^{V_2} PdV = nRT\ln(V_2/V_1) = nRT\ln(P_1/P_2).$$

We may also expand a gas *adiabatically* so that no heat is exchanged with the surroundings during the expansion. The temperature of the gas will not remain constant during such an adiabatic expansion. Now

$$dw = -PdV \qquad \text{(Section 2.1)}$$

and

$$dU = C_V dT \qquad \text{(Section 2.8).}$$

Also as

$$dq = 0, \qquad dU - dw = 0$$

and we have

$$C_V dT + PdV = 0.$$

Dividing by T and substituting $P = RT/V$ for one mole of perfect gas we find

$$C_V \frac{dT}{T} + R\frac{dV}{V} = 0.$$

If the gas is in an initial state T_1, V_1 and expands to a final state T_2, V_2 we can integrate this equation (if C_V is assumed to be independent of temperature) to obtain

$$C_V \ln(T_2/T_1) + R \ln(V_2/V_1) = 0.$$

As for one mole of gas

$$C_P - C_V = R \qquad \text{(Section 2.8)},$$

$$C_V \ln T_2/T_1 + (C_P - C_V) \ln V_2/V_1 = 0$$

and

$$\left(\frac{T_1}{T_2}\right) = \left(\frac{V_2}{V_1}\right)^{(C_P/C_V - 1)}.$$

As for a perfect gas the initial and final states must satisfy the perfect-gas equation of state

$$\frac{T_1}{T_2} = \frac{P_1 V_1}{P_2 V_2}$$

and

$$P_1 V_1^{C_P/C_V} = P_2 V_2^{C_P/C_V} = \text{constant}.$$

The change in internal energy for an adiabatic expansion of one mole of gas from V_1 to V_2 is calculated using the relation derived in Section 2.7;

$$\Delta U = C_V(T_2 - T_1).$$

The initial and final temperatures are related to the volumes by the equation derived above.

8.2 Irreversible expansion

If the pressure on a gas is suddenly lowered from a value P_1 to P_2 (Fig. 8.1) and the gas is allowed to expand against this new pressure to a volume V_2 it does work

$$w = -\int_{V_1}^{V_2} P \, dV = -P_2(V_2 - V_1).$$

If this expansion is carried out adiabatically so that $q = 0$, then, as $\Delta U = q + w$,

$$w = \Delta U = C_V(T_2 - T_1).$$

Equating these two expressions for w allows T_2 to be evaluated.

$$C_V(T_2 - T_1) = -P_2\left(\frac{nRT_2}{P_2} - \frac{nRT_1}{P_1}\right).$$

Then w may be evaluated by substituting into the expression for ΔU.

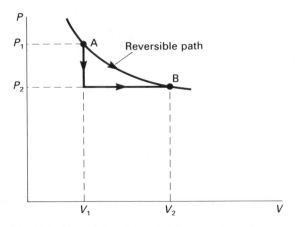

Fig. 8.1. Reversible and irreversible expansions of a gas.

Example

Let us consider the expansion of $0.01\ m^3$ of a perfect gas from a pressure of 1 atm to a pressure of 0.10 atm at 298 K. If the gas is monatomic the molar heat capacities are $C_P = \frac{5}{2}R$ and $C_V = \frac{3}{2}R$ (Section 9.4). The number of moles of gas is given by

$$n = \frac{PV}{RT} = \frac{1 \times 1.013 \times 10^5 \times 0.01}{8.314 \times 298}$$

$$n = 0.409 \text{ moles.}$$

(We note that the pressure must be converted to the unit $N\,m^{-2}$ to perform this calculation.)

Isothermal reversible expansion

As $P_1 V_1 = P_2 V_2$ for a perfect gas

$$V_2 = \frac{1}{0.10} \times 0.01 = 0.10\,m^3.$$

$$\text{Then } -w_{rev} = nRT \ln \frac{V_2}{V_1} \qquad \text{(Section 3.5)}$$

$$w_{rev} = -0.409 + 8.31 \times 298 \times \ln 10$$

$$w_{rev} = -2331\ J.$$

Adiabatic reversible expansion

As $P_1 V_1^{C_P/C_V} = P_2 V_2^{C_P/C_V}$ (Section 8.1) we have

$$V_2 = \left(\frac{P_1}{P_2}\right)^{C_V/C_P} V_1$$

$$V_2 = (10)^{3/5} \times 0.01$$
$$V_2 = 0.0398 \text{ m}^3.$$

The final temperature is now given by

$$T_2 = \frac{P_2 V_2}{nR} = \frac{0.1 \times 1.013 \times 10^5 \times 0.0398}{0.409 \times 8.314}$$

$$T_2 = 118.6 \text{ K}.$$

As, for this adiabatic process $q = 0$, we obtain $\Delta U = w = C_V(T_2 - T_1)$. Noting $C_V = n\frac{3}{2}R$ we obtain

$$w = 0.409 \times 1.5 \times 8.314(118.6 - 298)$$
$$w = -915 \text{ J}.$$

Irreversible adiabatic expansion

The work done is (Section 8.1) given by

$$w = -P_2(V_2 - V_1) = -P_2\left(\frac{nRT_2}{P_2} - \frac{nRT_1}{P_1}\right).$$

But as $q = 0$ we obtain $\Delta U = C_V(T_2 - T_1)$ and hence

$$C_V(T_2 - T_1) = -P_2\left(\frac{nRT_2}{P_2} - \frac{nRT_1}{P_1}\right)$$

Thus

$$1.5R(T_2 - 298) = RT_2 + R \times 298 \times \frac{0.1}{1.0}$$

which gives $T_2 = 190.7 \text{ K}$.

$$w = \Delta U = \tfrac{3}{2}nR(190.7 - 298)$$
$$w = -547.2 \text{ J}.$$

We note that the irreversible process does less work than the equivalent reversible process and the gas does not cool to the same extent.

8.3 Equation of state of gases

The variation of internal energy of a gas that does work only against atmospheric pressure is $dU = TdS - PdV$ (Section 3.4). We can write this equation in the form

$$\frac{dU}{dV} = T\left(\frac{dS}{dV}\right) - P.$$

For changes at constant temperature we may write the equation in the more restricted form

$$\left(\frac{\partial U}{\partial V}\right)_T = T\left(\frac{\partial S}{\partial V}\right)_T - P.$$

Recalling that $dA = dU - TdS - SdT$ (Section 4.1) and $dU = TdS - PdV$, we obtain $dA = -SdT - PdV$. We can see

$$\left(\frac{\partial A}{\partial T}\right)_V = -S \quad \text{and} \quad \left(\frac{\partial A}{\partial V}\right)_T = -P.$$

As the order of differentiation of a state function does not matter,

$$\frac{\partial}{\partial V}\left(\frac{\partial A}{\partial T}\right)_V = -\left(\frac{\partial S}{\partial V}\right)_T \quad \text{and} \quad \frac{\partial}{\partial T}\left(\frac{\partial A}{\partial V}\right)_T = -\left(\frac{\partial P}{\partial T}\right)_V,$$

i.e.

$$\left(\frac{\partial S}{\partial V}\right)_T = \left(\frac{\partial P}{\partial T}\right)_V.$$

This relation is one of a number known as *Maxwell's relations* whose usefulness is not confined to the problem at hand; they are valuable in a number of areas. Using this relation we obtain

$$\left(\frac{\partial U}{\partial V}\right)_T = T\left(\frac{\partial P}{\partial T}\right)_V - P,$$

an equation which relates the variation of energy with volume to the equation of state, that is the P, V, T relations of the gas (or indeed liquid or solid).
For one mole of a perfect gas $PV = RT$ (Section 1.5), and

$$\left(\frac{\partial P}{\partial T}\right)_V = \frac{R}{V}.$$

Therefore

$$\left(\frac{\partial U}{\partial V}\right)_T = \frac{RT}{V} - P = P - P = 0.$$

For a perfect gas $\left(\frac{\partial U}{\partial V}\right)_T = 0$: the internal energy is independent of volume. We have assumed earlier that this is so, in fact our definition of a perfect gas included this condition. Now we have shown it to be a consequence of the perfect-gas equation of state. Closely similar arguments show that for a perfect gas $\left(\frac{\partial H}{\partial P}\right)_T = 0.$

8.4 The Joule–Thomson experiment

In 1843 Joule showed, within the limits of error of his apparatus, that the expansion of gas into a vacuum, as in Fig. 8.2, was accompanied by no temperature change. As in this experiment $\mathrm{d}w = 0$, and since he observed that $\mathrm{d}q$ was also effectively zero,

$$\mathrm{d}U = \mathrm{d}q + \mathrm{d}w = 0 \qquad \text{(Section 2.5)}$$

and

$$\left(\frac{\partial U}{\partial V}\right)_T = 0.$$

More careful measurements would have shown that for real gases $\left(\frac{\partial U}{\partial V}\right)_T$ was not exactly zero.

A superior way of investigating these effects was devised by Joule and Thomson. In their apparatus, illustrated in Fig. 8.3, gas flowed at a steady rate through the porous plug. The system was insulated so that $\mathrm{d}q = 0$. The force applied to the piston forcing the gas through the plug is equal to the product of the pressure P_1 and the area of the piston a. The distance through which the point of application of the force moves is equal to the volume of the gas driven through the plug divided by the area. Thus when one mole of gas is moved,

$$\text{work done on gas} = \text{force} \times \text{distance}$$

$$= P_1 a\left(\frac{V_1}{a}\right) = P_1 V_1.$$

The net work is this work of compression less the work recovered when the gas expands on the far side of the plug. Thus

$$w = P_1 V_1 - P_2 V_2.$$

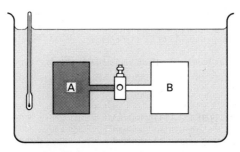

Fig. 8.2. The Joule experiment.

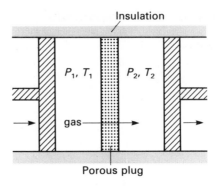

Fig. 8.3. The Joule–Thomson experiment.

If the gas on both sides of the plug followed the perfect-gas equation then w would be zero. As $\Delta U = q + w$ and $q = 0$ (Section 2.5),

$$\Delta U = U_2 - U_1 = P_1 V_1 - P_2 V_2.$$

Therefore

$$U_2 + P_2 V_2 = U_1 + P_1 V_1,$$

and from Section 2.7

$$\Delta H = H_2 - H_1 = U_2 + P_2 V_2 - U_1 - P_1 V_1,$$

or

$$\Delta H = 0.$$

Thus the Joule–Thomson experiment occurs at constant enthalpy. The Joule–Thomson coefficient is defined as $\left(\dfrac{\partial T}{\partial P}\right)_H$ and is determined by the change in temperature of the gas for a fixed pressure drop across the porous plug. The degree to which $\left(\dfrac{\partial T}{\partial P}\right)_H$ differs from zero provides an indication of the energy that arises from interactions between the molecules of a gas. The molecules of a perfect gas do not interact and as we have seen above for such a gas $\left(\dfrac{\partial T}{\partial P}\right)_H$ is zero as is $\left(\dfrac{\partial U}{\partial V}\right)_T$. Most of the common gases cool when passing from high to low pressures in a Joule–Thomson apparatus at room temperature. This has been used as a method of liquefying gases.

8.5 Imperfect gases: fugacity

So far we have assumed that the gases we have been dealing with are perfect. This is found to be a remarkably good approximation. For water vapour just above its boiling point the errors introduced in PV at 1 atm would be only 1.5 per cent. For nitrogen at 298 K and 10 atm pressure the volume calculated assuming perfect-gas behaviour would be in error by less than 0.5 per cent. A perfect gas will follow the equation

$$\mu = \mu^0 + RT \ln(P/\text{atm}) \qquad \text{(Section 4.10)}.$$

We may define a new thermodynamic function, the fugacity f such that for real gases

$$\mu = \mu^0 + RT \ln(f/\text{atm})$$

and

$$f/\text{atm} = \exp\left\{\frac{(\mu - \mu_0)}{RT}\right\}.$$

We could regard f as the 'effective pressure' just as activities were introduced as 'effective concentrations'. As all gases tend to perfect behaviour as $P \to 0$, under these conditions f will tend to become equal to P. The standard state is strictly the state of the perfect gas at one atmosphere pressure where $P = 1$ atm and $f = 1$ atm. This is illustrated in Fig. 8.4. For practical purposes, as many gases are virtually ideal at 1 atm, no great error is introduced if the standard state is characterized by the properties of the real gas at 1 atm.

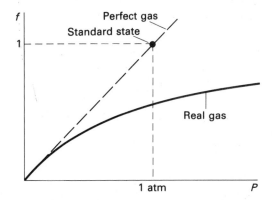

Fig. 8.4. Definition of the standard state of unit fugacity for an imperfect gas.

Recalling the calculations of gas equilibria carried out earlier, we have for imperfect gases

$$A(g) \rightleftharpoons B(g)$$

$$\mu_A = \mu_A^0 + RT\ln(f_A/\text{atm}), \qquad \mu_B = \mu_B^0 + RT\ln(f_B/\text{atm})$$

$$\Delta G' = \Delta G^0 + RT\ln(f_B/f_A),$$

$$K_f = (f_B/f_A)_{\text{equil}}$$

is the strict definition of equilibrium constant for gas reactions, and

$$\Delta G^0 = -RT\ln K_f.$$

8.6 Calculation of fugacities

As for 1 mole of pure gas $\mu = \mu^0 + RT\ln(f/\text{atm})$ and μ^0 is independent of pressure, we have for an imperfect gas

$$\left(\frac{\partial \mu}{\partial P}\right)_T = V = RT\left(\frac{\partial \ln f}{\partial P}\right)_T,$$

where V is the volume of a mole of the gas. If the equation of state of the gas is known the fugacity may be determined directly using

$$\left[\frac{\partial \ln(f/P)}{\partial P}\right]_T = \left[\frac{\partial \ln f}{\partial P}\right]_T - \left[\frac{\partial \ln P}{\partial P}\right]_T = \frac{V}{RT} - \frac{1}{P}$$

This equation may be integrated from a low pressure where the fugacity is equal to the pressure to a higher pressure P giving

$$\ln\frac{f}{P} = \int_0^P \left(\frac{V}{RT} - \frac{1}{P}\right) dP.$$

For gases that deviate only moderately from perfect-gas behaviour a more convenient approximate relation can be used:

$$\frac{f}{P} = \frac{P}{P_{\text{per}}}$$

where P_{per} is the pressure a perfect gas would have at the same volume.

Table 8.1 Fugacities of nitrogen at 273 K

Pressure/atm	f/P	P/P_{per}
1	0.99955	0.99955
10	0.9956	0.9957
100	0.9703	0.9851
1000	1.839	2.070

Fugacities calculated using this equation for nitrogen at 273 K are compared with the exact values in Table 8.1.

Problems

8.1. Calculate the final volume and the work done when $5 \times 10^{-3}\,m^3$ of a perfect (monatomic) gas at 273 K is reversibly expanded from 10 atm to 1 atm pressure (a) isothermally and (b) adiabatically. What is the final temperature after the adiabatic expansion? Assume $C_V = 3R/2$.

8.2. Calculate the final temperature and the work done if $5 \times 10^{-3}\,m^3$ of a perfect (monatomic) gas at 273 K and at 10 atm pressure is allowed to expand suddenly (irreversibly and adiabatically) against an external pressure of 1 atm.

8.3. Methane at 20 atm pressure and 223 K has a molar volume of $8.3 \times 10^{-4}\,m^3$. Estimate the fugacity of methane under these conditions.

9

The molecular basis of thermodynamics

As we have discovered, thermodynamics gives an account of the interrelation of *macroscopic* properties. It does not directly invoke the molecular nature of matter and its conclusions appear to be independent even of the existence of molecules. However, we can relate the conclusions of thermodynamics to the way in which molecules behave and the establishment of this link, statistical thermodynamics, is the subject of this chapter.

9.1 Energy levels

Quantum mechanics tells us that energies are 'quantized'—that only certain values of the energy are possible. For simple systems, we can calculate these permitted values—the *energy levels* of the system—using the basic equation of quantum theory, the Schrödinger equation. Each permitted state and its corresponding energy is defined in terms of *quantum numbers* which have integral, or, in special cases, half-integral values.

The simplest example is a particle confined to a box (Fig. 9.1). As the box is made smaller or the particles less massive the energy levels become further apart. For a cubic box the energy levels may be expressed as

$$\varepsilon = \frac{h^2}{8ml^2}(n_x^2 + n_y^2 + n_z^2),$$

where l is the length of the box, m the mass of the particle, h Planck's constant, and n_x, n_y, and n_z the quantum numbers for the motion in the x-, y-, and z-directions. Thus if a molecule of a gas is confined to a container the kinetic energy associated with the motion of its centre of mass (called its *translational energy*) is quantized.

Polyatomic molecules can also have rotational and vibrational energy and we can write (to a good approximation)

$$\varepsilon_{total} = \varepsilon_{trans} + \varepsilon_{rot} + \varepsilon_{vib}.$$

The rotational and vibrational energies are also quantized. The energy levels

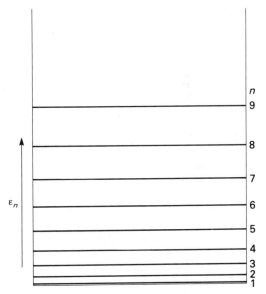

Fig. 9.1. Energy levels of a particle in a one-dimensional box.

associated with the rotation of a linear molecule are given by the expression

$$\varepsilon_{\text{rot}} = \left(\frac{h^2}{8\pi^2 I}\right) J(J + 1),$$

where J is the *rotational quantum number* and I the moment of inertia. $I = \Sigma m_i r_i^2$ for masses m_i a distance r_i from the centre of mass. For vibrational motion the energy levels are given by $\varepsilon_{\text{vib}} = (v + \frac{1}{2})hv$, where v is the vibrational quantum number and v the frequency of the vibration.

We see that even in its lowest energy level when $v = 0$ a vibrator has energy $\frac{1}{2}hv$. This is called its *zero-point* energy. Its existence is due to the uncertainty principle which is a fundamental consequence of quantum mechanics. It tells us that we cannot simultaneously specify exactly both the momentum and position of a particle. As particles are confined to a limited region of space (as will occur in vibrational motion) their linear momentum cannot be exactly zero and so the particles energy also cannot be zero. Hence an oscillator can never lose all its energy.

For a typical gas molecule under normal conditions the translational energy levels are so close together that we can regard translational energy as continuously variable. But this is not the case for rotational and vibrational levels. The spacing of rotational levels is of the order of $10 \, \text{J mol}^{-1}$ (though it can vary by more than a factor of ten from this value) (Table 9.1) and the vibrational level separation is of the order of $10 \, \text{kJ mol}^{-1}$. The pattern of energy levels that is often observed is illustrated in Fig. 9.2.

Table 9.1 Rotational and vibrational energy levels of diatomic molecules

$$\Delta\varepsilon_{rot} = \varepsilon_{J=1} - \varepsilon_{J=0}, \quad \Delta\varepsilon_{vib} = \varepsilon_{v=1} - \varepsilon_{v=0} = h\nu,$$
$$\theta_{rot} = (h^2/8\pi^2 I)/k, \quad \text{and} \quad \theta_{vib} = h\nu/k$$

	$\Delta\varepsilon_{rot}$ / J mol^{-1}	$2\theta_{rot}$ / K	$\Delta\varepsilon_{vib}$ / kJ mol^{-1}	θ_{vib} / K
H_2	1420	171	51.0	6140
HCl	254	30	35.8	4300
O_2	34.6	4.2	18.8	2260
N_2	48.1	5.8	28.1	3380
Cl_2	5.8	0.70	6.8	810
Br_2	1.9	0.23	4.0	470
I_2	0.90	0.11	2.6	310
CO	46.0	5.5	26.0	3120
NO	41.0	4.9	22.8	2740

Note: $2\theta_{rot}$ and θ_{vib} are those temperatures at which kT becomes equal to the energy corresponding to the first rotational and vibrational transitions respectively.

Fig. 9.2. Vibrational and rotational energy levels. v and J are the vibrational and rotational quantum numbers.

In order to be able to use our knowledge of energy levels to calculate thermodynamic properties we must find out how molecules distribute themselves among the available levels. A complication that arises is that different states, each an independent solution of the Schrödinger equation, may have the same energy. This energy level is then said to be *degenerate* with a degeneracy equal to the number of states that have this energy. Thus for each rotational energy ε_J of a linear molecule there are $(2J + 1)$ states (correspond-ing to the $(2J + 1)$ different orientations of the molecule that quantum theory permits). It is important when calculating the way molecules may distribute themselves among the energy levels to take account of degeneracy. It is often more convenient to discuss the problem in terms of states rather than energy levels. Each atom or molecule will be in one of the energy states. The total energy of a number of molecules

$$U = \sum_{\text{states}} \varepsilon_i n_i$$

depends on the energies of the states, ε_i, and the number of molecules that occupy them, n_i. Thus if we could find a rule that gave us n_i we could, from a knowledge of the energy levels and their degeneracies calculate the total energy of our system. More generally it turns out that a knowledge of ε_i and n_i serves to define all thermodynamic properties.

9.2 Microstates

We have already met this concept in Section 3.10 where we identified 'microstates' or 'complexions' as the number of ways the overall state of a system can be constructed. The number of ways a state of the system can arise determines the probability of its occurrence. If we toss two coins the probability of getting one head and one tail is twice that of getting two heads. If we write out all the possibilities (HT, TH, HH, TT) we see that this is because two 'microstates' give rise to the mixed configuration. Consider a fixed number of molecules distributed over energy levels. As a concrete example let us take three molecules distributed between equally spaced energy levels (Fig. 9.3) with the total energy fixed at three units. We see that three distinct distributions, I, II, and III are possible. But not all these distributions are equally probable. We can see why this is so if we assume the molecules can be individually identified and label each one so as to enumer-ate the microstates which go to make up each of the three different distribu-tions. We find that six times as many microstates make up II as III, and thus it is six times as probable (Fig. 9.3).

The general formula for the number of microstates, W, which make up a distribution is the same as that for the number of ways N objects can be distributed between a number of boxes so that there are n_1 in the first box, n_2 in the second and so on,

$$W = \frac{N!}{n_1! n_2! \ldots}.$$

The total number of possible arrangements is $N!$ but we must divide this by $n_1! n_2!$ etc. as we cannot distinguish between states that differ only by an exchange of particles within a given energy level. Thus for distribution I we

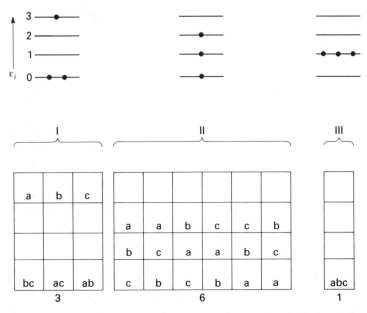

Fig. 9.3. Arrangements of three molecules amongst energy states (with the total energy restricted to three units).

get† $W_I = \dfrac{3!}{1!2!} = 3$, for $W_{II} = \dfrac{3!}{1!1!1!} = 6$, and for $W_{III} = \dfrac{3!}{3!} = 1$ as we found by direct enumeration. The most probable distribution of a large number of objects between a number of boxes, if we have no other constraints, is a uniform distribution as this would correspond to the maximum number of microstates. However, when we consider the distribution of the molecules over energy levels we wish to know how they will be distributed when the system is at thermodynamic equilibrium. This leads to a different type of distribution being favoured which is a compromise between a wide distribution over many states, tending to maximize W, and the restriction on total energy.

9.3 The Boltzmann factor

As discussed in Section 3.10, Boltzmann recognized that there was a close link between the entropy of a system and the number of microstates that comprise it which he expressed

$$S = k \ln W,$$

where k is Boltzmann's constant. Thus for a number of molecules distributed

† Note that $0! = 1$ and $1! = 1$.

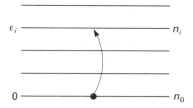

Fig. 9.4. One molecule moved from the lowest to the ith energy state. n_0 and n_i are the number of molecules in the two states before the move.

between energy levels

$$S = k \ln \left\{ \frac{N!}{n_1! n_2! \ldots} \right\}.$$

Consider N, a large number of molecules distributed between a set of energy states so that n_0 are in the lowest state and n_i in the ith. We shall assume that the numbers of molecules in each energy state are large. Let us examine the consequences of adding a small quantity of energy, ε_i, to the system to promote a molecule from the lowest state to the ith state (Fig. 9.4). The system we have chosen is a constant volume system and does no work.†
Thus when the system is at equilibrium, $dA = dU - TdS = 0$ (Section 4.1) and $dS = dU/T$. As the addition of ε_i represents a very small change in the energy of this large system the associated entropy change is given by

$$dS = \frac{dU}{T} = \frac{\varepsilon_i}{T}.$$

As we have seen in the previous section the entropy change can also be expressed in terms of the number of microstates that make up the system. The small change in entropy, dS, that results from moving a molecule from the 0th to the ith state in a very large system is given by

$$dS = k \ln \left\{ \frac{N!}{(n_0 - 1)! n_1! \ldots (n_i + 1)! \ldots} \right\} - k \ln \left\{ \frac{N!}{n_0! n_1! \ldots n_i! \ldots} \right\}$$

$$= -k \ln \left\{ \frac{(n_i + 1)!}{n_i!} \cdot \frac{(n_0 - 1)!}{n_0!} \right\}.$$

Thus

$$dS = -k \ln \left\{ \frac{(n_i + 1)}{n_0} \right\} \approx -k \ln \frac{n_i}{n_0}$$

† It can be shown that a system in which the energy levels remain unchanged (even when energy is added or subtracted) is one that does no PV work.

as $n_i \gg 1$. Equating the two expressions for dS we obtain

$$dS = \frac{\varepsilon_i}{T} = -k \ln \frac{n_i}{n_0}$$

or

$$\frac{n_i}{n_0} = e^{-\varepsilon_i/kT}. \dagger$$

We call $e^{-\varepsilon_i/kT}$ the *Boltzmann factor* of the ith energy state and the equation the *Boltzmann distribution*. It is one of the most important relations in physical science and provides great insight into the systems we deal with in physical chemistry.

The Boltzmann factor tells us that at low temperatures all molecules will tend to lie in the lowest possible state whereas at very high temperatures the distribution over the energy states tends to be uniform (Fig. 9.5). The energy of a system tends to zero at the absolute zero of temperature and, as all particles then lie in the lowest state, the number of complexions will be unity and $S = k \ln W = 0$. This is the basis for the Third Law of Thermodynamics (Section 5.7). We can write the Boltzmann distribution in terms of the probability of an energy state being occupied. If the total number of molecules is N then the probability of any particular molecule occupying the

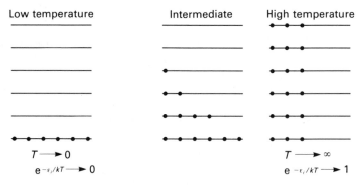

Fig. 9.5. The distribution of molecules among available energy states at various temperatures.

---

† Many books reach this result by considering systems in which the energy U (as well as N and V) is fixed. They consider small changes in occupation numbers n_i which conserve energy and for which, with this constraint, W is a maximum. The result is then derived using the method of undetermined multipliers (see Gasser and Richards 1974).

ith state, p_i, is

$$p_i = \frac{n_i}{N} = \frac{n_i/n_0}{N/n_0} = \frac{e^{-\varepsilon_i/kT}}{\Sigma e^{-\varepsilon_i/kT}}$$

as $N = \Sigma n_i$.

Although we have obtained the Boltzmann distribution law for systems in which the values of n_i are large, its validity is more general and it can be shown to apply equally well to systems in which the probability of any particular energy state being occupied is very small.

9.4 The behaviour of heat capacity

The behaviour of the heat capacities of substances both in the solid and gaseous phases was a mystery to scientists before the discovery of quantum theory. In classical mechanics where energy is regarded as continuously variable it can be shown that the energy of a system is divided equally between the various modes of motion called *degrees of freedom*. Furthermore, according to classical physics each degree of freedom contributed $RT/2$ to the total molar energy and $R/2$ (as $C_V = \left(\dfrac{dU}{dT}\right)_V$) to the molar heat capacity. Thus

$$\varepsilon_{trans} = \tfrac{1}{2}m\langle v_x^2 \rangle + \tfrac{1}{2}m\langle v_y^2 \rangle + \tfrac{1}{2}m\langle v_z^2 \rangle = \tfrac{3}{2}RT,$$

where v_x, v_y, v_z are the three components of molecular velocity and $\langle v_x^2 \rangle^{1/2}$ the root mean square velocity. $\varepsilon_{rot} = RT$ for diatomic and linear molecules with two modes of rotation and $\varepsilon_{rot} = \tfrac{3}{2}RT$ for non-linear polyatomic molecules. $\varepsilon_{vib} = RT$ for each vibration (each vibration gives two contributions of $\tfrac{1}{2}RT$ as both potential and kinetic energy terms contribute). Thus we can, using these classical results, predict the heat capacities of gases as shown in Table 9.2.

The classical result is correct for monatomic gases but considerably overestimates the heat capacities of diatomic and polyatomic gases, and

Table 9.2 The classical molar heat capacity of gases

	U_{trans}	U_{rot}	U_{vib}	U_{total}	C_V	C_P	$\gamma = C_P/C_V$
Monatomic	$3/2RT$	—	—	$3/2RT$	$3/2R$	$5/2R$	1.667
Diatomic	$3/2RT$	RT	RT	$7/2RT$	$7/2R$	$9/2R$	1.29
Polyatomic	$3/2RT$	$3/2RT$	xRT	$(3+x)RT$	$(3+x)R$	$(4+x)R$	$\left(\dfrac{4+x}{3+x}\right) \to 1$ as $x \to \infty$

x is the number of vibrations of molecule. For non-linear polyatomic molecules $x = 3N - 6$ and for linear molecules $x = 3N - 5$, where N is the number of atoms in the molecule. (Note: $C_P - C_V = R$.)

underestimates the value of the specific heat ratio, $\gamma = \dfrac{C_P}{C_V}$. However, the existence of discrete energy levels enables us to understand this failure. If energy levels are very far apart compared with kT then almost all molecules will be in the lowest level as $e^{-\varepsilon_i/kT} \to 0$. If we increase the temperature of the system slightly they will *still* be in the lowest level and the system will have absorbed no energy. Thus

$$C_V = \left(\frac{dU}{dT}\right)_V = 0$$

and at low temperatures the heat capacities of substances tend to zero. On the other hand if energy levels are close together compared with kT we can show that each degree of freedom contributes $R/2$ to the molar heat capacity as predicted by classical mechanics. At the temperatures at which kT is of the same order of magnitude as the separation between the energy levels there is a rapid increase in C_V. Translational motion is always classical since the separation between translational energy levels is always very much less than kT (at least at temperatures greater than 10^{-17} K). However, we can see that whether a rotational or vibrational mode contributes to the heat capacity depends on the temperatures and the separation of the energy levels (Fig. 9.6). For most molecules at room temperature (other than some molecules with very low moments of inertia such as H_2, HBr, CH_4, etc.) rotation contributes its full classical value to the heat capacity as $\Delta\varepsilon_{rot} < kT$ (see Table 9.1). (At

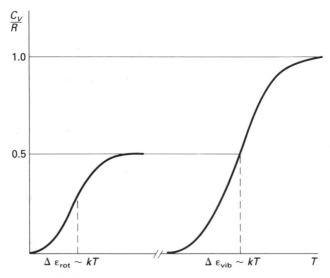

Fig. 9.6. The contribution to heat capacity of a vibrational and rotational degree of freedom.

300 K the value of RT is approximately $2.5 \, \text{kJ mol}^{-1}$.) However $\Delta\varepsilon_{\text{vib}}$ is usually much greater than kT and under these circumstances vibrations do not contribute significantly to heat capacity (Fig. 9.7). Thus for nitrogen at room temperature the rotational contribution is approximately RT and the vibrational contribution almost zero. Thus $C_V \approx (5/2)R$ and $\gamma \approx 7/5 = 1.40$ as opposed to the classical prediction of 1.29. For iodine the vibrational spacings are closer (Table 9.1) and we would predict $\gamma \approx 1.39$ in accord with the classical value. If the temperature is varied the heat capacity of a diatomic or polyatomic gas may show 'steps' as the contributions from rotations and vibrations rise as the energy separations become comparable to kT. The positions of the steps depend on the moments of inertia and the vibrational frequencies of the molecules.

The atoms in a monatomic solid can be regarded as three-dimensional oscillators. At high temperatures they contribute the classical value to the molar heat capacity and $C_V = 3R$ or about $25 \, \text{J K}^{-1} \, \text{mol}^{-1}$. The observation that the heat capacities of most monatomic solids have this value near room temperature is known as *Dulong and Petit's Law*. At low temperatures the heat capacity is zero, the temperature dependence having the same shape as

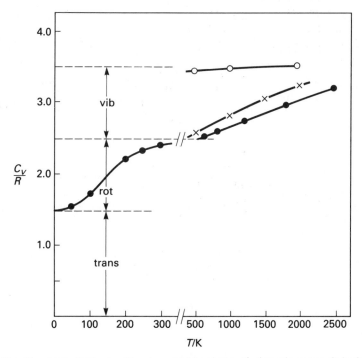

Fig. 9.7. The molar heat capacity at constant volume of diatomic gases ● hydrogen, × nitrogen, ○ iodine.

the vibrational heat capacity function illustrated in Fig. 9.6. This result for the heat capacity of solids was first obtained by Einstein. A more realistic theory due to Debye takes account of the fact that the atoms vibrate with more than one frequency.

9.5 Partition functions

We have found that the number of molecules in a specified state is given by

$$\frac{n_i}{n_0} = e^{-\varepsilon_i/kT} \qquad \text{(Section 9.3)}.$$

The average energy per molecule is given by

$$\frac{U}{N} = \frac{\Sigma \varepsilon_i n_i}{\Sigma n_i} = \frac{\Sigma \varepsilon_i n_i / n_0}{\Sigma n_i / n_0} = \frac{\Sigma \varepsilon_i e^{-\varepsilon_i/kT}}{\Sigma e^{-\varepsilon_i/kT}}.$$

It is convenient to identify the denominator as the *partition function, z*.

$$z = \sum_i e^{-\varepsilon_i/kT}.$$

We can see that z is a number related to the sum of the probabilities of the various energy states being occupied. We can regard it as a measure of the *effective number of energy states* accessible to the system. It will be larger for systems with energy levels close together and, for any particular system, it becomes larger as the temperature is increased. As discussed earlier many energy levels are *degenerate*, that is a number of different states of the molecule have the same energy. If the number of states contributing to the ith level is g_i (called the degeneracy of the ith level), then the partition function should be written in terms of energy levels as

$$z = \sum_{\text{states}} e^{-\varepsilon_i/kT} = \sum_{\text{levels}} g_i e^{-\varepsilon_i/kT}.$$

However, we shall continue to write the partition function in its simple form remembering that it represents a sum over *states* rather than energy levels.

The internal energy of a system can be expressed in terms of the partition function. We note

$$\left(\frac{\partial z}{\partial T}\right)_V = \left[\frac{\partial(\Sigma e^{-\varepsilon_i/kT})}{\partial T}\right]_V$$

$$= \frac{1}{kT^2} \Sigma \varepsilon_i e^{-\varepsilon_i/kT}.$$

Thus

$$U = \frac{N \Sigma \varepsilon_i e^{-\varepsilon_i/kT}}{\Sigma e^{-\varepsilon_i/kT}} = \frac{NkT^2 \left(\frac{\partial z}{\partial T}\right)_V}{z}$$

and

$$U = NkT^2 \left(\frac{\partial \ln z}{\partial T} \right)_V.$$

For one mole of substance N is equal to the Avogadro constant and as $N_A k = R$,

$$U = RT^2 \left(\frac{\partial \ln z}{\partial T} \right)_V.$$

Thus the internal energy of a molecular system may be expressed directly in terms of the partition function z. If the energy levels of the system are known, z, and hence U, can be evaluated. Thus a thermodynamic function can be calculated from a knowledge of molecular properties.

We shall see that just as the total energy of a molecular system can be regarded as arising from a number of distinct contributions from translational motion, rotation, and vibration,

$$\varepsilon = \varepsilon_{trans} + \varepsilon_{rot} + \varepsilon_{vib},$$

so we can write separate partition functions for each type of energy as $z = \Sigma e^{-\varepsilon_i/kT}$

and

$$\varepsilon_i = \varepsilon_{i_{trans}} + \varepsilon_{i_{rot}} + \varepsilon_{i_{vib}}.$$

Therefore

$$z = \Sigma \exp[-(\varepsilon_{i_{trans}} + \varepsilon_{i_{rot}} + \varepsilon_{i_{vib}})/kT]$$

which gives

$$z = \Sigma \exp(-\varepsilon_{i_{trans}}/kT) \Sigma \exp(-\varepsilon_{i_{rot}}/kT) \Sigma \exp(-\varepsilon_{i_{vib}}/kT)$$

and

$$z = z_{trans} \cdot z_{rot} \cdot z_{vib}.$$

In later sections we shall see how the partition functions for translational, rotational, and vibrational energy levels can be calculated.

9.6 Entropy and the partition function

The entropy of a system may be expressed following Boltzmann as

$$S = k \ln W.$$

We have seen that for N distinguishable particles distributed over a set of energy levels we obtain for the number of microstates, W,

$$W = \frac{N!}{n_1! n_2! \ldots}$$

and

$$\ln W = \ln N! - \Sigma \ln n_i!$$

If N and n_i are large we can apply Stirling's approximation

$$\ln N! \approx N \ln N - N.$$

Then $\ln W = N \ln N - N - (\Sigma n_i \ln n_i - \Sigma n_i)$, but $\Sigma n_i = N$ and thus $\ln W = - \Sigma n_i \ln \frac{n_i}{N}$, hence

$$S = - Nk\Sigma \frac{n_i}{N} \ln \frac{n_i}{N} = - Nk\Sigma p_i \ln p_i$$

where

$$p_i = \frac{n_i}{N}, \text{ the probability that the } i\text{th state is occupied.}$$

Now

$$p_i = \frac{e^{-\varepsilon_i/kT}}{z} \qquad \text{(Section 9.3)}$$

and

$$\ln p_i = - \frac{\varepsilon_i}{kT} - \ln z.$$

Substituting in the expression for entropy above we obtain

$$S = - Nk \left[- \frac{1}{kT} \Sigma p_i \varepsilon_i - \ln z \Sigma p_i \right].$$

But

$$\Sigma p_i = 1 \quad \text{and} \quad \Sigma p_i \varepsilon_i = \frac{U}{N},$$

thus

$$S = - Nk \left[- \frac{U}{NkT} - \ln z \right]$$

$$S = + \frac{U}{T} + Nk \ln z.$$

For one mole

$$S = \frac{U}{T} + R \ln z,$$

or from the expression for U in terms of the partition function z given in Section 9.5,

$$S = R \ln z + RT \left(\frac{\partial \ln z}{\partial T} \right)_V.$$

Again we find that a thermodynamic function S can be expressed in terms of z, which itself can be calculated from molecular properties.

These expressions however are restricted to solids, as our expression for the number of microstates assumed that the molecules were in some way identifiable. Now this is true in solids where their positions on lattice sites can be specified but is not true in a gas where an exchange of molecules does not produce a distinguishable state. We must, in the case of perfect gases, divide the number of microstates by $N!$ to allow for this possibility of interchange.† This leads to an additional term in the expression for entropy

$$- k \ln N! = - kN \ln N + Nk.$$

Thus we obtain for perfect gases

$$S = \frac{U}{T} + R \ln z - R \ln N_A + R$$

or

$$S = R \ln z + RT \left(\frac{\partial \ln z}{\partial T} \right)_V - R \ln N_A + R.$$

As $A = U - TS$ the molar Helmholtz free energy for a perfect gas becomes

$$A = - RT \ln z + RT \ln N_A - RT$$

$$A = - RT \ln \left(\frac{z}{N_A} \right) - RT.$$

---

† This would lead to $W = \dfrac{N!}{n_i! n_2! \dots N!} \cdot \dfrac{1}{N!} = \dfrac{1}{n_1! n_2! \dots}$ and to values of W less than unity. This paradox can only be resolved by performing a calculation which takes account of the fact that there may be many more states than molecules. See G. N. Lewis and M. Randall (1961). *Thermodynamics;* revised by K. S. Pitzer and L. Brewer; Appendix 3. McGraw-Hill, N.Y.

The Gibbs free energy for one mole of perfect gas may be written

$$G = A + PV = A + RT$$

$$G = -RT\ln\left(\frac{z}{N_A}\right).$$

Other thermodynamic functions can be obtained in terms of the partition function using standard thermodynamic formulae (Table 9.3). Note that the terms that arise from the $N!$ which modify the expressions for S, A, and G for gases are associated only with the translational contribution to the partition function. When evaluating the contributions of vibration and rotation we use the simple expression as given for solids in Table 9.3.

Table 9.3 Molar thermodynamic functions in terms of the molecular partition function, z

Solids	Perfect gases
$U = RT^2\left(\dfrac{\partial \ln z}{\partial T}\right)_V$	$U = RT^2\left(\dfrac{\partial \ln z}{\partial T}\right)_V$
$S = R\ln z + RT\left(\dfrac{\partial \ln z}{\partial T}\right)_V$	$S = R\ln z + RT\left(\dfrac{\partial \ln z}{\partial T}\right)_V - R\ln N_A + R$
$A = U - TS = -RT\ln z$	$A = -RT\ln z + RT\ln N_A - RT$
$P = -\left(\dfrac{\partial A}{\partial V}\right)_V = RT\left(\dfrac{\partial \ln z}{\partial V}\right)_T$	$P = RT\left(\dfrac{\partial \ln z}{\partial V}\right)_T$
$H = U + PV$	
$= RT\left[T\left(\dfrac{\partial \ln z}{\partial T}\right)_V + V\left(\dfrac{\partial \ln z}{\partial V}\right)_T\right]$	$H = RT\left[T\left(\dfrac{\partial \ln z}{\partial T}\right)_V + V\left(\dfrac{\partial \ln z}{\partial V}\right)_T\right]$
$G = H - TS$	$G = RT\left[\ln z - V\left(\dfrac{\partial \ln z}{\partial V}\right)_T - \ln N_A + 1\right]$
$= -RT\left[\ln z - V\left(\dfrac{\partial \ln z}{\partial V}\right)_T\right]$	$= -RT\ln\left(\dfrac{z}{N_A}\right)$
$C_V = \left(\dfrac{\partial U}{\partial T}\right)_V = RT\left[2\left(\dfrac{\partial \ln z}{\partial T}\right)_V + T\left(\dfrac{\partial^2 \ln z}{\partial T^2}\right)_V\right]$	$C_V = RT\left[2\left(\dfrac{\partial \ln z}{\partial T}\right)_V + T\left(\dfrac{\partial^2 \ln z}{\partial T^2}\right)_V\right]$
$\mu_i = -RT\ln(z_i e^{-Y/RT})$	$\mu_i = -RT\ln\left(\dfrac{z_i}{N_i}\right)$

where Y is the interaction energy of the atoms or molecules in the solid.

It is also possible to express the thermodynamic properties in terms of what is called a *canonical partition function*, Z. This is related to z by $Z = z^N$ for an ideal solid and by $Z = z^N/N!$ for a perfect gas, where z is the molecular partition function.

Then the relations

$$U = kT^2 \left(\frac{\partial \ln Z}{\partial T} \right)_V$$

and

$$S = k \ln Z + kT \left(\frac{\partial \ln Z}{\partial T} \right)_V$$

apply to both states of matter.

9.7 Calculation of the translational partition function

The energy levels of a molecule confined to a cubic box of side l are given, as indicated in Section 9.1, by

$$\varepsilon_{\text{trans}} = \frac{n_x^2 h^2}{8ml^2} + \frac{n_y^2 h^2}{8ml^2} + \frac{n_z^2 h^2}{8ml^2},$$

the terms arising from motion in the x-, y-, and z-directions; m is the mass of the molecule. Considering only the x-direction we may write the partition function

$$z_{x_{\text{trans}}} = \sum_{n_x = 1}^{\infty} \exp\left(-\frac{n_x^2 h^2}{8ml^2 kT} \right).$$

However, the translational levels are so close together we can replace the summation by an integral over n_x

$$z_{x_{\text{trans}}} = \int_0^{\infty} \exp\left(-\frac{n_x^2 h^2}{8ml^2 kT} \right) dn_x,$$

which we can write in the form $I = \displaystyle\int_0^{\infty} e^{-an^2} \, dn$, where $a = \left(\dfrac{h^2}{8ml^2 kT} \right)$. This is a standard integral $I = \dfrac{1}{2}\left(\dfrac{\pi}{a} \right)^{1/2}$ so we obtain

$$z_{x_{\text{trans}}} = \frac{\pi^{1/2}}{2} \left(\frac{8ml^2 kT}{h^2} \right)^{1/2} = \left(\frac{2\pi mkT}{h^2} \right)^{1/2} l.$$

An identical contribution arises from motion in the y- and z-directions and

$$z_{\text{trans}} = z_{x_{\text{trans}}} \cdot z_{y_{\text{trans}}} \cdot z_{z_{\text{trans}}},$$

giving

$$z_{trans} = \left(\frac{2\pi mk T}{h^2}\right)^{3/2} l^3 = \left(\frac{2\pi mk T}{h^2}\right)^{3/2} V,$$

where the volume of the cubic container, V, is equal to l^3. Evaluating the constants in this expression we obtain

$$\ln z_{trans} = \frac{3}{2}\ln\left(\frac{m}{a.m.u.}\right) + \frac{5}{2}\ln\left(\frac{T}{K}\right) - \ln\left(\frac{P}{atm}\right) + 51.104;$$

a.m.u. is the atomic mass unit which corresponds to 1/12 of the mass of a carbon atom. The numerical value of $(m/a.m.u.)$ is equal to the molecular weight (relative molecular mass). For one mole of gaseous argon (relative molecular mass = 40) at 1 atmosphere pressure and 298 K we find

$$z_{trans} = 6.1 \times 10^{30}.$$

The translational partition functions of gases are very large which tells us that there are many translational energy levels accessible to the molecules at normal temperatures. We can use the partition function to calculate the contribution of translational motion to the molar thermodynamic properties employing the relations of Table 9.3.

Thus $U = RT^2\left(\dfrac{\partial \ln z}{\partial T}\right)_V = RT^2 \dfrac{3}{2} \cdot \dfrac{1}{T} = \dfrac{3}{2} RT$, the classical value

$$S = R\ln z + RT\left(\frac{\partial \ln z}{\partial T}\right)_V - R\ln N_A + R$$

$$\frac{S_{trans}}{R} = \frac{3}{2}\ln\left(\frac{2\pi mk T}{h^2}\right) + \ln V + \frac{3}{2} - \ln N_A + 1.$$

This equation is known as the Sackur–Tetrode equation. We obtain for one mole

$$\frac{S_{trans}}{R} = \ln\left(\frac{V}{m^3}\right) + \frac{3}{2}\ln\left(\frac{T}{K}\right) + \frac{3}{2}\ln\left(\frac{m}{a.m.u.}\right) + 8.23,$$

where the constant is $\dfrac{5}{2} + \dfrac{3}{2}\ln\left(\dfrac{2\pi k}{h^2}\right) - \ln N_A + \dfrac{3}{2}\ln\left(\dfrac{a.m.u.}{kg}\right)$. In terms of pressure we obtain

$$\frac{S_{trans}}{R} = \frac{3}{2}\ln\left(\frac{m}{a.m.u.}\right) + \frac{5}{2}\ln\left(\frac{T}{K}\right) - \ln\left(\frac{P}{atm}\right) - 1.1650.$$

Substituting the values for argon at 1 atm and 298 K leads to a molar entropy of 154.8 J K^{-1} mol^{-1}, in excellent agreement with the value obtained by calorimetry (Sections 5.7, 5.8). As $P = RT\left(\dfrac{\partial \ln z}{\partial V}\right)_T$ we obtain

$$P = RT\dfrac{\partial \ln\left[\left(\dfrac{2\pi mk T}{h^2}\right)^{3/2} V\right]}{\partial V} = \dfrac{RT}{V}.$$

Thus as we would expect the gas follows the perfect-gas equation of state.

9.8 The rotational partition function

We stated earlier (Section 9.1) that for the rotational motion of a linear molecule the energy levels are given by

$$\varepsilon_{\text{rot}} = \dfrac{J(J+1)h^2}{8\pi^2 I}.$$

Each rotational level has a degeneracy of $(2J + 1)$ so there are $(2J + 1)$ *states* at each level. For most molecules other than hydrogen the rotational energy levels are sufficiently close compared with RT (see Table 9.1) that in order to evaluate the partition functions we can replace the summation by an integration over J.

$$z_{\text{rot}} = \int_0^\infty (2J+1) \exp\left[\dfrac{-J(J+1)h^2}{8\pi^2 IkT}\right] dJ = \dfrac{8\pi^2 IkT}{h^2}.$$

This is the result for a *heteronuclear* diatomic molecule. In order to generalize it we must include a factor σ that takes into account that for a homonuclear diatomic, A_2, a full rotation gives rise to two indistinguishable orientations $A - A'$ and $A' - A$. This reduces the number of different terms contributing to the partition function by two. Thus

$$z_{\text{rot}} = \dfrac{8\pi^2 IkT}{h^2 \sigma}.$$

σ is called the *symmetry number* and is unity for a heteronuclear diatomic (or asymmetrical linear molecule) and 2 for a homonuclear diatomic or symmetrical linear molecule. Evaluating the constants leads to the expression

$$\ln z_{\text{rot}} = \ln\left(\dfrac{I}{\text{a.m.u. nm}^2}\right) + \ln\left(\dfrac{T}{K}\right) - \ln \sigma + 1.418.$$

More generally for polyatomic molecules with three rotations we obtain

$$z_{\text{rot}} = \dfrac{8\pi^2 (8\pi^2 I_x I_y I_z)^{1/2}(kT)^{3/2}}{\sigma h^3},$$

where I_x, I_y, and I_z are the moments of inertia about three mutually perpendicular axes and σ is again the symmetry number defined more generally as the number of indistinguishable positions which the molecule can attain by rotation (assuming we cannot distinguish between different atoms of the same species in the same position). For water $\sigma = 2$, for ammonia 3, and for methane 12. (Three identical configurations of methane can be obtained by rotation about each of the four carbon–hydrogen bonds.) At 298 K we find z_{rot} for nitrogen is 52 and its value for chlorine is 417. Hydrogen is unusual in having rotational levels exceptionally far apart so that only very few are accessible at room temperature, and for this substance $z_{rot} = 1.8$. For most linear molecules U_{rot} at room temperature has the classical value RT as suggested by classical mechanics for two degrees of rotational motion. Using the expression in Table 9.3, the entropy is given by

$$S_{rot}/R = 1 + \ln\left(\frac{8\pi^2 IkT}{h^2\sigma}\right)$$

for diatomic and linear molecules. This can be expressed

$$S_{rot}/R = \ln\left(\frac{I}{\text{a.m.u. nm}^2}\right) + \ln\left(\frac{T}{K}\right) - \ln\sigma + 2.418.$$

For non-linear polyatomics we obtain

$$S_{rot}/R = \tfrac{1}{2}\ln\left(\frac{I_x I_y I_z}{\text{a.m.u.}^3\,\text{nm}^6}\right) + \frac{3}{2}\ln\left(\frac{T}{K}\right) - \ln\sigma + 4.197.$$

9.9 Vibrational partition function

The energy levels of a molecule vibrating with simple harmonic motion are given by $\varepsilon_{vib} = hv(v + \tfrac{1}{2})$, where v is the vibrational frequency and v the vibrational quantum number. To obtain the vibrational partition function we sum the energy levels measuring the energy from the lowest available level, $\varepsilon_0 = \tfrac{1}{2}hv$ (the zero-point energy), to obtain

$$z_{vib} = \sum_{v=0}^{v=\infty} \exp\left(\frac{-hvv}{kT}\right) = 1 + \exp\left(\frac{-hv}{kT}\right) + \exp\left(\frac{-2hv}{kT}\right)$$
$$+ \exp\left(\frac{-3hv}{kT}\right) + \ldots\ .$$

This is the geometric series $1 + x + x^2 + x^3 \ldots$ for which the sum to infinity for $x < 1$ is $(1 - x)^{-1}$. Thus

$$z_{vib} = \left[1 - \exp\left(\frac{-hv}{kT}\right)\right]^{-1}.$$

Vibrational partition functions are usually close to unity as $\dfrac{hv}{kT} > 1$, indicating that only the lowest vibrational level is accessible at normal temperatures. Using the relations of Table 9.3 we can find expressions for the (molar) thermodynamic functions in terms of hv/kT.

Thus $\dfrac{U - U_0}{RT} = \dfrac{x}{e^x - 1}$, where $x = \dfrac{hv}{kT}$ and U_0 is the molar zero-point energy $U_0 = \frac{1}{2}Nhv$,

$$\frac{C_V}{R} = \frac{x^2 e^x}{(e^x - 1)^2} \quad \text{and} \quad \frac{S}{R} = x(e^x - 1)^{-1} - \ln(1 - e^{-x}).$$

These functions are tabulated in Table 9.4. On the whole vibrations make only a small contribution to the thermodynamic properties except in the case of very slow vibrations which can occur in polyatomic molecules.

9.10 Evaluation of the thermodynamic properties of gaseous nitrogen

We shall now see the usefulness of statistical thermodynamics by calculating the thermodynamic properties of gaseous nitrogen at 298 K. The translational entropy of a gas at 1 atm can be calculated from the equations in Section 9.7.

$$S_{\text{trans}}/R = \ln\left(\frac{V}{\text{m}^3}\right) + \frac{3}{2}\ln\left(\frac{T}{\text{K}}\right) + \frac{3}{2}\ln\left(\frac{m}{\text{a.m.u.}}\right) + 8.23.$$

The relative molecular mass of nitrogen is 28 and

$$S_{\text{trans}}/R = \ln(0.0244) + \tfrac{3}{2}\ln(298) + \tfrac{3}{2}\ln(28) + 8.23$$

$$S_{\text{trans}}/R = 18.07$$

and so $S_{\text{trans}} = 150.2 \ \text{J K}^{-1}\,\text{mol}^{-1}$.

Table 9.4 Thermodynamic properties of a harmonic oscillator

$\dfrac{hv}{kT}$	C_V/R	$\dfrac{U - U_0}{RT}$	$-\left(\dfrac{G - U_0}{RT}\right)$	S/R
0.00	1.000	1.000	∞	∞
0.10	0.999	0.951	2.352	3.303
0.50	0.979	0.771	0.933	1.704
1.00	0.921	0.582	0.459	1.041
2.00	0.724	0.313	0.145	0.458
3.00	0.469	0.157	0.051	0.208
5.00	0.171	0.034	0.007	0.041
10.00	0.005	0.000	0.000	0.000

U_0 is the molar zero-point energy of the vibrational motion, $U_0 = \frac{1}{2}Nhv$.

To calculate the rotational entropy we need to know the bond length, which is 0.109 nm. Thus the moment of inertia

$$I = 2 \times 14 \times \left(\frac{0.109}{2}\right)^2 = 0.0832 \text{ a.m.u. nm}^2.$$

Then, using the expression obtained in Section 9.8,

$$S_{rot}/R = \ln\left(\frac{I}{\text{a.m.u. nm}^2}\right) + \ln\left(\frac{T}{\text{K}}\right) - \ln \sigma + 2.418$$

$$S_{rot}/R = \ln(0.0832) + \ln(298) - \ln(2) + 2.418$$

$$= 4.94$$

$$S_{rot} = 41.1 \text{ J K}^{-1}\text{mol}^{-1}.$$

To evaluate the contribution from vibrational motion we need the vibrational frequency for nitrogen. This is reported by spectroscopists as 2360 cm^{-1}. This unit, which is favoured in spectroscopy, is that of the actual frequency divided by the speed of light.

To obtain the actual frequency we must multiply the value by $3 \times 10^{10} \text{ cm s}^{-1}$, which gives us $7.08 \times 10^{13} \text{ s}^{-1}$.

Thus

$$\frac{h\nu}{kT} = \frac{6.62 \times 10^{-34} \times 7.08 \times 10^{13}}{1.38 \times 10^{-23} \times 298} = 11.4$$

and inspection of Table 9.4 shows that the vibrational contribution to entropy is negligible.

The total molar entropy for nitrogen at 1 atm and 298 K is therefore

$$S = 150.2 + 41.1 = 191.3 \text{ J K}^{-1}\text{mol}^{-1}.$$

The contribution to the internal energy (measured relative to the zero-point energy) and to the heat capacity of the vibrational motion is also negligible (Table 9.4) whereas that from translational and rotational motion is classical. Thus

$$U = \tfrac{3}{2}RT + RT = \tfrac{5}{2}RT$$

and $C_V = \tfrac{5}{2}R$.

9.11 Chemical equilibrium

Some of the most valuable results we have obtained from chemical thermodynamics are those that relate the position of chemical equilibrium to the thermodynamic properties of the reactants and products. With the aid of statistical thermodynamics we can go one step further and relate the position of equilibrium to the masses, dimensions, and vibrational frequencies of the molecules involved.

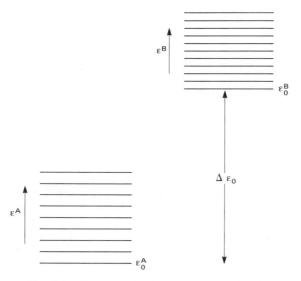

Fig. 9.8. Energy levels of two substances A and B.

Consider the equilibrium between two chemical species A and B. Each possesses a series of energy states, the energies of which we assign ε_j^A and ε_k^B when measured from the lowest levels of each species. That is ε_k^B is the energy of the kth state of B measured from the lowest level of B, ε_0^B, and ε_j^A is the energy of the jth level of A measured from ε_0^A (see Fig. 9.8).

The number of molecules in any of the states of A or B is given by

$$n_i = n_0 e^{-\varepsilon_i/kT},$$

where n_0 is the number of molecules in the lowest energy level of the whole system (ε_0^A as drawn in Fig. 9.7). The numbers of molecules in each form will be given by

$$n_A = \sum_j n_j^A = n_0 \sum_j \exp(-\varepsilon_j^A/kT)$$

and

$$n_B = \sum_k n_k^B - n_0 \sum_k \exp[-(\varepsilon_k^B + \Delta\varepsilon_0)/kT],$$

where $\Delta\varepsilon_0 = \varepsilon_0^B - \varepsilon_0^A$, the difference between the lowest energy levels of the two forms. Thus

$$\frac{n_B}{n_A} = \frac{\sum_k \exp(-\varepsilon_k^B/kT)}{\sum_j \exp(-\varepsilon_j^A/kT)} \cdot \exp(-\Delta\varepsilon_0/kT).$$

The two sums are the partition functions of the two species and the equilibrium ratio n_B/n_A can be identified as an equilibrium constant K thus:

$$K = \frac{n_B}{n_A} = \frac{z_B}{z_A} \exp(-\Delta\varepsilon_0/kT)$$

We can also define other equilibrium constants in terms of partition functions. For an equilibrium

$$aA + bB \rightleftharpoons lL + mM,$$

we can define an equilibrium constant in terms of partial pressures

$$K_P = \frac{(P_L/atm)^l(P_M/atm)^m}{(P_A/atm)^a(P_B/atm)^b} \qquad \text{(Section 4.13)}.$$

Now $-RT \ln K_P = \Delta G^0$ (Section 4.11), where ΔG^0 is the change in the Gibbs free energy for a mole of reaction with all reactants and products in their standard states at a pressure of 1 atm.

As $G = -RT \ln \left(\dfrac{z}{N_A}\right)$ (Section 9.6)

$$\ln K_P = -(lG_L^0 + mG_M^0 - aG_A^0 - bG_B^0)/RT$$

$$K_P = \frac{\left(\dfrac{z_L^0}{N_A}\right)^l \left(\dfrac{z_M^0}{N_A}\right)^m}{\left(\dfrac{z_A^0}{N_A}\right)^a \left(\dfrac{z_B^0}{N_A}\right)^b} \exp(-\Delta\varepsilon_0/kT),$$

where z_i^0 are the molecular partition functions evaluated for the standard state, $P = 1$ atm.

It can also be shown that K_c, the equilibrium constant in terms of concentrations, is related to the partition functions evaluated per unit volume

$$K_c = \frac{\left(\dfrac{z_L}{V}\right)^l \left(\dfrac{z_M}{V}\right)^m}{\left(\dfrac{z_A}{V}\right)^a \left(\dfrac{z_B}{V}\right)^b} \cdot \exp(-\Delta\varepsilon_0/kT).$$

Example

Calculate the equilibrium constant K_P for the equilibrium

$$I_2(g) \rightleftharpoons 2I(g)$$

at 1173 K. The dissociation energy of I_2 is 1.544 eV, the vibrational wave number is 214.5 cm^{-1}, and the moment of inertia 4.51 a.m.u. nm^2. The ground state of the iodine atom has a degeneracy of 4 and the relative atomic mass of the iodine atom is 126.9.

We can use the relation obtained above in the form

$$K_P = \left[\frac{z^0(\mathrm{I})}{N_A} \right]^2 \bigg/ \frac{z^0(\mathrm{I}_2)}{N_A}.$$

Using the factorization of the partition function given in Section 9.5 we obtain

$$K_P = \frac{[z^0_{\mathrm{trans}}(\mathrm{I})]^2 \exp[-(\Delta\varepsilon_0/kT)]}{N_A[z^0_{\mathrm{trans}}(\mathrm{I}_2) \cdot z_{\mathrm{rot}}(\mathrm{I}_2) \cdot z_{\mathrm{vib}}(\mathrm{I}_2)]}.$$

Thus

$$\ln K_P = 2 \ln [z^0_{\mathrm{trans}}(\mathrm{I})] - \ln N_A - \ln [z^0_{\mathrm{trans}}(\mathrm{I}_2)]$$

$$- \ln [z_{\mathrm{rot}}(\mathrm{I}_2)] - \ln [z_{\mathrm{vib}}(\mathrm{I}_2)] - \frac{\Delta\varepsilon_0}{kT}.$$

We now evaluate each of these terms.

Translational partition function for I atoms: to the relation for z_{trans} given in Section 9.7 we must add the contribution from the degeneracy $\ln g$

$$\ln z_{\mathrm{trans}} = \frac{3}{2} \ln \left(\frac{m}{\mathrm{a.m.u.}} \right) + \frac{5}{2} \ln \left(\frac{T}{K} \right) - \ln \left(\frac{P}{\mathrm{atm}} \right) + 51.104 + \ln g$$

$$= \tfrac{3}{2} \ln (126.9) + \tfrac{5}{2} \ln (1173) - \ln (1) + 51.104 + \ln 4$$

$$\ln z^0_{\mathrm{trans}}(\mathrm{I}) = \underline{77.423}.$$

Translational partition function for I_2: using the same equation we obtain

$$\ln z_{\mathrm{trans}} = \tfrac{3}{2} \ln (253.8) + \tfrac{5}{2} \ln (1173) - \ln (1) + 51.104$$

$$\ln z^0_{\mathrm{trans}}(\mathrm{I}_2) = \underline{77.07.}$$

Rotational partition function for I_2: substituting into the equation of Section 9.8 we obtain

$$\ln z_{\mathrm{rot}} = \ln \left(\frac{I}{\mathrm{a.m.u.\ nm}^2} \right) + \ln \left(\frac{T}{K} \right) - \ln \sigma + 1.418$$

$$\ln z_{\mathrm{rot}}(\mathrm{I}_2) = \ln (4.51) + \ln (1173) - \ln 2 + 1.418$$

$$\ln z_{\mathrm{rot}}(\mathrm{I}_2) = \underline{9.298.}$$

Vibrational partition function of I_2: the unit given for the vibrational wave number as in the earlier example (Section 9.10) is a reciprocal wavelength $(1/\lambda)$ and is related to a frequency by $\nu = c/\lambda$, where c is the velocity of light. The relevant parameter x (see Section 9.9) is given by

$$x = \frac{hc(1/\lambda)}{kT} = \frac{6.626 \times 10^{-34} \times 3 \times 10^8 \times 214.5 \times 10^2}{1.380 \times 10^{-23} \times 1173}$$

$$x = 0.263$$

$$\ln z_{\mathrm{vib}} = \ln (1 - e^{-0.263})^{-1} \qquad \text{(Section 9.9)}$$

$$\ln z_{\mathrm{vib}}(\mathrm{I}_2) = \underline{1.464}.$$

The term involving the difference in energies is given by $\dfrac{\Delta\varepsilon_0}{kT}$. $\Delta\varepsilon_0$ is 1.544 eV which is 2.474 × 10^{-19} J (see endpaper for conversion factors).

$$\frac{\Delta\varepsilon_0}{kT} = \frac{2.474 \times 10^{-19}}{1.381 \times 10^{-23} \times 1173} = \underline{15.272} \, .$$

The sum of these terms, together with

$$\ln N_A = \underline{54.755}$$

gives

$$\ln K_P = -3.020$$

and

$$K_P = 0.0488.$$

Two experimental determinations by different workers give K_P at 1173 K as 0.0474 and 0.0481, in excellent agreement.

Problems

9.1. Calculate the translational partition function and entropy of one mole of xenon at 1 atm pressure and 298 K. The relative molecular mass of xenon is 131.30.

9.2. Using the data given in Section 9.11 calculate the equilibrium constant K_P for the reaction $I_2 \rightleftharpoons 2I$ at 973 K.

9.3. Calculate the translational, rotational, and vibrational contributions to the entropy of HCl at 1 atm and 298 K.

$$\frac{\theta_{rot}}{K} = 15 \qquad \frac{\theta_{vib}}{K} = 4300$$

and the relative molecular masses of Cl and H are 35.45 and 1.01 respectively.

Answers to problems

2.1 0.11 K
2.2 153 s
2.3 0.1 J
2.4 3 kJ
2.5 76 J K^{-1}

3.1 -9.4 J K^{-1}
3.2 28.7 J K^{-1}
3.3 109 J K^{-1}
3.4 0.33
3.5 34.7 J K^{-1}
3.6 -13.7 J K^{-1}

4.1 3.5 K
4.2 57.7 kJ mol^{-1}, 122.6 J K^{-1} mol^{-1}, 12.0 kJ mol^{-1}
4.3 -58.8 kJ per mole of dimer
4.4 143 kJ
4.5 -12.6 kJ
4.6 0.84 atm
4.7 63.4 kJ mol^{-1}, 105.5 J K^{-1} mol^{-1}, 34.6 K J mol^{-1}, 0 (the boiling point of mercury is found to be 600 K)
4.8 92.3 kJ mol^{-1}, 16.2 J K mol^{-1}, 9.4 kJ mol^{-1}

5.1 -3261 kJ mol^{-1} (after conversion of ΔU to ΔH), 42 kJ
5.2 -136 kJ mol^{-1}
5.3 -239 kJ mol^{-1}
5.4 38.3 kJ mol^{-1}
5.5 70.7 J K^{-1} mol^{-1}
5.6 0.28 atm, 0.50 atm

6.1 0.51, 0.49, 0.30, 0.70
6.2 116, or 124 if approximate formulae are used
6.3 3.8
6.4 9.4 kJ
6.5 114 000
6.6 0.22

7.1 0.82, 0.56, 1.0, 2.8
7.2 10^{37}
7.3 2×10^{-10}
7.4 -204 kJ, -112 kJ

8.1 (a) 50×10^{-3} m^3, -11.6 kJ
     (b) 19.9×10^{-3} m^3, -4.6 kJ, 109 K
8.2 175 K, -2.7 kJ
8.3 17.9 atm

9.1 3.6×10^{31}, 169.6 J K^{-1}
9.2 1.9×10^{-3}
9.3 153.6 J K^{-1} mol^{-1}, 33.2 J K^{-1} mol^{-1}, negligible

Appendix 1

Thermochemical data at 298.15 K

The thermodynamic quantities listed are for one mole of substance in its standard state, that is at 1 atm pressure. The enthalpies and free energies of formation of substances are the changes in those thermodynamic properties when a substance in its standard state is formed from its elements in their standard states. The standard state of an element is its normal physical state at 1 atm, and for the data given in these tables, 298.15 K. The entropies listed are 'absolute' in the sense that they are based on the assumption that the entropy of a pure substance is zero at the absolute zero of temperature.

Substance	ΔH_f^o kJ mol^{-1}	ΔG_f^o kJ mol^{-1}	S^o J K^{-1} mol^{-1}	C_P^o J K^{-1} mol^{-1}
Ag(s)	0.00	0.00	42.701	25.48
AgBr(s)	− 99.49	− 95.939	107.1	52.38
AgCl(s)	− 127.03	− 109.72	96.10	50.79
AgI(s)	− 62.38	− 66.31	114.	54.43
Al(s)	0.00	0.00	28.32	24.33
Al$_2$O$_3$(s)	− 1669.7	− 1576.4	50.986	78.99
Ar(g)	0.00	0.00	154.7	20.786
Br(g)	111.7	82.38	174.912	20.786
Br$_2$(g)	30.7	3.1421	248.48	35.9
Br$_2$(l)	0.00	0.00	152.0	
C(g)	718.384	672.975	157.992	20.837
C(diamond)	1.8961	2.8660	2.4388	60.62
C(graphite)	0.00	0.00	5.6940	86.44
CCl$_4$(g)	− 106.0	− 64.0	309.4	392.9
CH$_4$(g)	− 74.847	− 50.793	186.1	35.71
CO(g)	− 110.523	− 137.268	197.90	29.14
CO$_2$(g)	− 393.512	− 394.383	213.63	37.12
C$_2$H$_2$(g)	226.747	209.20	200.81	43.927
C$_2$H$_4$(g)	52.283	68.123	219.4	43.55
C$_2$H$_6$(g)	− 84.667	− 32.88	229.4	52.655
C$_3$H$_8$(g)	− 103.8			
n-C$_4$H$_{10}$(g)	− 126.16	− 17.15	309.9	97.45
i-C$_4$H$_{10}$(g)	− 134.53	− 2.92	294.6	96.82
CH$_3$OH(l)	− 283.6	− 166.3	126.7	81.6
CCl$_4$(l)	− 139.5	− 68.74	214.4	131.8
CS$_2$(l)	+ 87.9	+ 63.6	151.0	75.5

Thermochemical data at 298.15 K.—contd.

Substance	ΔH_f^0 kJ mol^{-1}	ΔG_f^0 kJ mol^{-1}	S^0 J K^{-1} mol^{-1}	C_P^0 J K^{-1} mol^{-1}
$C_2H_5OH(l)$	− 227.63	− 174.8	161.0	111.5
$CH_3CO_2H(l)$	− 487.0	− 392.5	159.8	123.4
$C_6H_6(l)$	49.028	172.8	124.50	
$Ca(s)$	0.00	0.00	41.6	26.2
$CaCO_3(calcite)$	− 1206.8	− 1128.7	92.8	81.88
$CaCO_3(aragonite)$	− 1207.0	− 1127.7	88.7	81.25
$CaC_2(s)$	− 62.7	− 67.7	70.2	62.34
$CaCl_2(s)$	− 794.6	− 750.1	113.0	72.63
$CaO(s)$	− 635.5	− 604.1	39.0	42.80
$Ca(OH)_2(s)$	− 986.58	− 896.75	76.1	84.5
$Cl(g)$	121.38	105.40	165.087	21.841
$Cl_2(g)$	0.00	0.00	222.94	33.9
$Cu(s)$	0.00	0.00	33.3	24.46
$CuCl(s)$	− 134	− 118	91.6	
$CuCl_2(s)$	− 205			
$CuO(s)$	− 155	− 127	43.5	44.3
$Cu_2O(s)$	− 166.6	− 146.3	100	69.8
$Fe(s)$	0.00	0.00	27.1	25.2
$Fe_2O_3(s)$	− 822.1	− 740.9	89.9	104.0
$Fe_3O_4(s)$	− 1117.0	− 1014	146	
$H(g)$	217.94	203.23	114.611	20.786
$HBr(g)$	− 36.2	53.22	198.47	29.1
$HCl(g)$	− 92.311	− 95.265	186.67	29.1
$HI(g)$	25.9	1.29	206.32	29.1
$H_2(g)$	0.00	0.00	130.58	28.83
$H_2O(g)$	− 241.826	− 228.595	188.72	33.57
$H_2O(l)$	− 285.840	− 237.191	69.939	75.295
$H_2S(g)$	− 20.14	− 33.02	205.6	33.9
$Hg(g)$	60.83	31.7	174.8	20.78
$Hg(l)$	0.00	0.00	77.4	27.8
$HgCl_2(s)$	− 230			76.6
$HgO(s, red)$	− 90.70	− 58.534	71.9	45.73
$HgO(s, yellow)$	− 90.20	− 58.404	73.2	
$Hg_2Cl_2(s)$	− 264.9	− 210.66	195.0	101
$I(g)$	106.61	70.148	180.682	20.786
$I_2(g)$	62.24	19.37	260.57	36.8
$I_2(s)$	0.00	0.00	116	54.97
$K(s)$	0.00	0.00	63.5	29.1
$KBr(s)$	392.1	− 379.1	96.44	53.63
$KCl(s)$	− 435.868	− 408.32	82.67	51.50
$KI(s)$	− 327.6	− 322.2	104.3	55.06
$Mg(s)$	0.00	0.00	32.5	23.8
$MgCl_2(s)$	− 641.82	− 592.32	89.5	71.29
$MgC(s)$	− 601.82	− 569.56	26.0	37.4
$Mg(OH)_2(s)$	− 924.66	− 833.74	63.13	77.02
$N(g)$	358.00	340.87	153.195	20.786
$NH_2(g)$	− 46.19	− 16.63	192.5	35.66
$NO(g)$	90.374	86.688	210.61	29.86

Thermochemical data at 298.15 K.—contd.

Substance	$\Delta H_f^{\ominus}$ kJ mol^{-1}	$\Delta G_f^{\ominus}$ kJ mol^{-1}	$S^{\ominus}$ J K^{-1} mol^{-1}	$C_P^{\ominus}$ J K^{-1} mol^{-1}
$NO_2(g)$	33.85	51.839	240.4	37.9
$N_2(g)$	0.00	0.00	191.48	29.12
$N_2O(g)$	81.54	103.5	219.9	38.70
$N_2O_4(g)$	9.660	98.286	304.3	79.07
$Na(s)$	0.00	0.00	51.0	28.4
$NaBr(s)$	− 359.94			52.3
$NaCl(s)$	− 411.00	− 384.02	72.38	49.70
$NaHCO_3(s)$	− 947.6	− 851.8	102.0	87.61
$NaOH(s)$	− 426.72			80.3
$Na_2CO_3(s)$	− 1130	− 1047	135	110.4
$O(g)$	247.52	230.09	160.953	21.909
$O_2(g)$	0.00	0.00	205.02	29.35
$Pb(s)$	0.00	0.00	64.89	26.8
$PbCl_2(s)$	− 359.1	− 313.9	136	76.9
$PbO(s, yellow)$	− 217.8	− 188.4	69.4	48.53
$PbO_2(s)$	− 276.6	− 218.9	76.5	64.4
$Pb_3O_4(s)$	− 734.7	− 617.5	211	147.0
$S(s, rhombic)$	0.00	0.00	31.8	22.5
$S(s, monoclinic)$	0.029	0.096	32.5	23.6
$SO_2(g)$	− 296.8	− 300.3	248.5	39.78
$SO_3(g)$	− 395.1	− 370.3	256.2	50.62
$S_8(g)$	100			
$Si(s)$	0.00	0.00	18.7	19.8
$SiO_2(s, quartz)$	− 859.3	− 805.0	41.84	44.43
$Zn(s)$	0.00	0.00	41.6	25.0
$ZnCl_2(s)$	− 415.8	− 369.25	108	76.5
$ZnO(s)$	− 347.9	− 318.1	43.9	40.2

A set of key thermodynamic values has been accepted by international agreement. These are given in *Journal of Chemical Thermodynamics*, **3**, 1 (1971).

Appendix 2

Thermodynamic data for ions in aqueous solution at 298.15 K

The standard state adopted for ions in aqueous solution is an ideal solution of unit molality. The values of thermodynamic properties listed are based on the assumption that the values for H^+ in such a solution are zero.

Ion	$\Delta H_f^{\ominus}$ (kJ mol^{-1})	$\Delta G_f^{\ominus}$ (kJ mol^{-1})	$S^{\ominus}$ (J K^{-1} mol^{-1})	$C_P^{\ominus}$ (J K^{-1} mol^{-1})
$Ag^+(aq)$	105.9	77.111	73.93	38
$Ba^{++}(aq)$	− 538.36	− 560.7	13	—
$Br^-(aq)$	− 120.9	− 102.82	80.71	− 128.4
$Ca^{++}(aq)$	− 542.96	− 553.04	− 55.2	—
$Cl^-(aq)$	167.45	131.17	55.10	− 125.5
$H^+(aq)$	[0.00]	[0.00]	[0.00]	[0.00]
$I^-(aq)$	55.94	51.67	109.4	130.0
$K^+(aq)$	− 251.2	− 282.28	102.5	—
$Mg^{++}(aq)$	− 461.96	− 456.01	− 118.0	—
$Na^+(aq)$	− 239.66	− 261.87	60.2	—
$NH_4^+(aq)$	− 132.8	− 79.50	112.0	—
$OH^-(aq)$	− 229.94	− 157.30	− 105.4	− 134.0
$S^{--}(aq)$	42	84	—	—
$SO_4^{--}(aq)$	− 907.51	− 741.99	17.2	17
$SO_3^{--}(aq)$	− 624.3	− 532.2	43.5	—
$Zn^{++}(aq)$	− 152.4	− 147.21	− 106.5	—

Further reading

Elementary texts

Mahan, B. H. (1963) *Elementary chemical thermodynamics*. Benjamin, New York. A very lucid introduction to the subject more elementary than this book.
Nash, L. K. (1962) *Elements of chemical thermodynamics*. Addison-Wesley, Reading, Massachusetts. Another concise introduction to the subject.

More advanced text books

Atkins, P. W. (1990) *Physical chemistry*, fourth edition, Oxford University Press, Oxford.
Bent, H. A. (1965) *The second law*. Oxford University Press, New York. An exciting and interesting approach. The best bed-time reading in chemical thermodynamics with fascinating historical quotations.
Caldin, E. F. (1958) *An introduction to chemical thermodynamics*. Oxford University Press, Oxford. A moderately advanced textbook with very full explanations of difficult points.
Everett, D. H. (1971) *Chemical thermodynamics*. Longman, London. A book with an original and stimulating approach to the subject. The nomenclature, though highly systematic, sometimes proves difficult for beginners in the subject.
Lewis, G. N. and Randall, M. revised Pitzer, K. S. and Brewer, L. (1961) *Thermodynamics*. McGraw-Hill, N.Y. A lucid, thorough, and comprehensive treatment of the subject.

Related Oxford Chemistry Series books

Pass, G. (1973) *Ions in solution* (3). Clarendon Press, Oxford.
Gasser, R. P. H. and Richards, W. G. (1974) *Entropy and energy levels*. Clarendon Press, Oxford.

1A	IIA	IIIA	IVA	VA	VIA	VIIA	VIII			IB	IIB	IIIB	IVB	VB	VIB	VIIB	O
1H 1·008																	2He 4·003
3Li 6·941	4Be 9·012											5B 10·81	6C 12·01	7N 14·01	8O 16·00	9F 19·00	10Ne 20·18
11Na 22·99	12Mg 24·31											13Al 26·98	14Si 28·09	15P 30·97	16S 32·06	17Cl 35·45	18Ar 39·95
19K 39·10	20Ca 40·08	21Sc 44·96	22Ti 47·90	23V 50·94	24Cr 52·00	25Mn 54·94	26Fe 55·85	27Co 58·93	28Ni 58·71	29Cu 63·55	30Zn 65·37	31Ga 69·72	32Ge 72·59	33As 74·92	34Se 78·96	35Br 79·90	36Kr 83·80
37Rb 85·47	38Sr 87·62	39Y 88·91	40Zr 91·22	41Nb 92·91	42Mo 95·94	43Te 98·91	44Ru 101·1	45Rh 102·9	46Pd 106·4	47Ag 107·9	48Cd 112·4	49In 114·8	50Sn 118·7	51Sb 121·8	52Te 127·6	53I 126·9	54Xe 131·3
55Cs 132·9	56Ba 137·3	57La 138·9	72Hf 178·5	73Ta 180·9	74W 183·9	75Re 186·2	76Os 190·2	77Ir 192·2	78Pt 195·1	79Au 197·0	80Hg 200·6	81Ti 204·4	82Pb 207·2	83Bi 209·0	84Po (210)	85At (210)	86Rn (222)
87Fr (223)	88Ra 226·0	89Ac (227)															

Lanthanides	57La 138·9	58Ce 140·1	59Pr 140·9	60Nd 144·2	61Pm (147)	62Sm 150·4	63Eu 152·0	64Gd 157·3	65Tb 158·9	66Dy 162·5	67Ho 164·9	68Er 167·3	69Tm 168·9	70Yb 173·0	71Lu 175·0
Actinides	89Ac (227)	90Th 232·00	91Pa 231·0	92U 238·0	93Np 237·0	94Pu (242)	95Am (243)	96Cm (248)	97Bk (247)	98Cf (251)	99Es (254)	100Fm (253)	101Md (256)	102No (254)	103Lw (257)

SI units

Physical quantity	Old unit	Value in SI units
energy	calorie (thermochemical)	4.184 J (joule)
	*electronvolt—eV	1.602×10^{-19} J
	*electronvolt per molecule	96.48 kJ mol^{-1}
	erg	10^{-7} J
	*wave number—cm^{-1}	1.986×10^{-23} J
entropy (S)	eu = cal g^{-1} °C^{-1}	4184 J kg^{-1} K^{-1}
force	dyne	10^{-5} N (newton)
pressure (P)	atmosphere	1.013×10^{5} Pa (pascal), or N m^{-2}
	torr = mmHg	133.3 Pa
dipole moment (μ)	debye—D	3.334×10^{-30} C m
magnetic flux density (H)	*gauss—G	10^{-4} T (tesla)
frequency (v)	cycle per second	1 Hz (hertz)
relative permittivity (ε)	dielectric constant	1
temperature (T)	*°C and °K	1 K (kelvin); 0 °C = 273.2 K

(* indicates permitted non-SI unit)

Multiples of the base units are illustrated by length

fraction	10^9	10^6	10^3	1	(10^{-2})	10^{-3}	10^{-6}	10^{-9}	(10^{-10})	10^{-12}
prefix	giga-	mega-	kilo-	metre	(centi-)	milli-	micro-	nano-	(*ångstrom)	pico-
unit	Gm	Mm	km	m	(cm)	mm	μm	nm	(*Å)	pm

The fundamental constants

Avogadro constant	L or N_A	6.022×10^{23} mol^{-1}
Bohr magneton	μ_B	9.274×10^{-24} J T^{-1}
Bohr radius	a_0	5.292×10^{-11} m
Boltzmann constant	k	1.381×10^{-23} J K^{-1}
charge of a proton	e	1.602×10^{-19} C
(charge of an electron = $-e$)		
Faraday constant	F	9.649×10^{4} C mol^{-1}
gas constant	R	8.314 J K^{-1} mol^{-1}
nuclear magneton	μ_N	5.051×10^{-27} J T^{-1}
permeability of a vacuum	μ_0	$4\pi \times 10^{-7}$ H m^{-1}, or N A^{-2}
permittivity of a vacuum	ε_0	8.854×10^{-12} F m^{-1}
Planck constant	h	6.626×10^{-34} J s
(Planck constant)/2π	$\hbar$	1.055×10^{-34} J s
rest mass of electron	m_e	9.110×10^{-31} kg
rest mass of proton	m_p	1.673×10^{-27} kg
speed of light in a vacuum	c	2.998×10^{8} m s^{-1}

$\ln 10 = 2.303$ $\ln x = 3.303 \lg x$ $\lg e = 0.4343$ $\pi = 3.142$

$R \ln 10 = 19.14$ J K^{-1} mol^{-1} $RTF^{-1} \ln 10 = 59.16$ mV at 298.2 K

Index